Palgrave Studies in Economic History

Series Editor
Kent Deng
London School of Economics
London, UK

Palgrave Studies in Economic History is designed to illuminate and enrich our understanding of economies and economic phenomena of the past. The series covers a vast range of topics including financial history, labour history, development economics, commercialisation, urbanisation, industrialisation, modernisation, globalisation, and changes in world economic orders.

More information about this series at
http://www.palgrave.com/gp/series/14632

Kristin Ranestad

Knowledge-Based Growth in Natural Resource Intensive Economies

Mining, Knowledge Development and Innovation in Norway 1860–1940

palgrave
macmillan

Kristin Ranestad
University of Oslo
Oslo, Norway

Palgrave Studies in Economic History
ISBN 978-3-030-40454-3 ISBN 978-3-319-96412-6 (eBook)
https://doi.org/10.1007/978-3-319-96412-6

Library of Congress Control Number: 2018951787

Cover illustration: ivind Eide / EyeEm

This Palgrave Macmillan imprint is published by the registered company Springer Nature Switzerland AG
The registered company address is: Gewerbestrasse 11, 6330 Cham, Switzerland

Acknowledgements

First, I would like to express my very great appreciation to my thesis supervisor, Kristine Bruland, who helped and guided me through my initial research into this topic and then through the whole process of writing this book. Without her suggestions, advice, and assistance, this research would not have been possible, and I am very grateful to her. I could not have had a more encouraging and helpful mentor.

This research has been implemented with the help of many people and I am thankful to all of them. I would like to offer my special thanks to my colleagues at the Paul Bairoch Institute of Economic History at the University of Geneva, especially Sylvain Wenger, who has helped me a lot. I want to give a special recognition to Keith Smith, Olav Wicken, and Mary O'Sullivan for very valuable comments and suggestions. Archivists in Chile, Norway, and the United States have been very helpful in the search for, collection of, and copy of records. At the University Library and State Archive in Trondheim, I would like to thank the archivists who helped me with the records from Røros Copper Works and Trondhjem Mechanical Workshop. The project "Sustainable Development, Fiscal Policy and Natural Resources Management. Bolivia, Chile and Peru in the Nordic countries' mirror" has contributed to the completion of this book. I want to thank my siblings and parents, who were a mental support through this process. My mother, Tone, my father, Per, and Karen and Erling also gave me useful comments along the way. I appreciate the support from all my friends very much, especially my very good friend

Linn Brandt who encouraged and accompanied me through long hours and nights of work. Finally, I want to thank my boyfriend, Paul Sharp for his support, suggestions, and being wonderful.

Contents

About the Author

Kristin Ranestad holds a PhD in economic history from the University of Geneva. Her thesis is a comparative study of Chile and Norway, two "natural resource–intensive economies", which have had divergent economic development trajectories, yet are closely similar in industrial structure and geophysical conditions. Ranestad is currently doing a post-doc at the University of Oslo and participating in the research project "Copper in the early modern period. A comparative study of work and everyday life in Falun and Røros". The project explores copper production in Røros and Falun with the aim of exploring the two copper mines from a global perspective and connecting them to global markets. She started a postdoc position at Lund University in April 2018 where she uses Scandinavian biographies to explore (1) the direct roles of formal education and practice for innovation and economic performance, and how this played out in different industries and (2) access to education and the historical impact of education on social mobility, and differences in this regard between genders and regions.

List of Figures

List of Tables

Part I

Theoretical and Historiographical Framework

1

Introduction

Do Natural Resources Lead to Slow Growth?

Economies in which natural resource sectors account for at least 10 per cent of gross domestic product (GDP), a share of export of at least 20–40 per cent, or where such sectors represent "key stone" sectors, have been defined in the literature as "natural resource–intensive economies".[1] Their natural resource industries rest essentially on production of raw materials, such as agriculture, forestry, and extraction of metals and minerals, which make them different from countries which base their economies on manufacturing or high-tech industries. Nobel Laureate in economics Douglass C. North defined "industrialised" societies as "region[s] whose export base consists primarily of finished consumers' goods and/or finished manufactured producers' goods".[2] The economist Keith Smith describes natural resource–intensive economies as countries

[1] S. Ville and O. Wicken, "The Dynamics of Resource-Based Economic Development: Evidence from Australia and Norway", *Industrial and Corporate Change*, Volume 22, Issue 5, (1 October 2013), p. 14.

[2] D. North, "Location Theory and Regional Economic Growth", *Journal of Political Economy* 63 (1955), p. 254.

K. Ranestad, *Knowledge-Based Growth in Natural Resource Intensive Economies*, Palgrave Studies in Economic History, https://doi.org/10.1007/978-3-319-96412-6_1

> with a strong emphasis on agriculture, a small manufacturing sector with a large proportion of output concentrated in low and medium-technology sectors, and a large service sector incorporating a large social and community services element. [...] Natural resources may provide a significant proportion of output, but more commonly a large proportion of exports. [...] Significant natural resources may include agricultural land, timber and forests, fish, hard rock minerals, and oil and gas.[3]

Some of the poorest countries in the world fit this description, notably African and Latin American countries. The poor economic performance of these countries has led to the notion that natural resources directly cause slow growth and retard development. An important argument for this is that natural resources "crowd out" manufacturing. Imports are therefore focused on manufactures, which means that countries which base their economies on natural resource industries find themselves in a vicious circle, or a "resource curse", in which they are always in external deficit because of declining terms of trade between natural resources and manufactures. These features prevent the economic progress that characterises industrialised countries and are understood to be important reasons why natural resource–intensive economies show poor economic performance.

Multiple negative symptoms are categorised under the resource curse. From the 1950s, the "dependency theory" had major effects on economic policy in Latin America. It stated that resources flowed from peripheral underdeveloped countries to the core of industrialised and developed countries. The idea was that developed countries became rich at the expense of underdeveloped countries. There were nuances within this approach, but they had some common traits. The economist Matias Vernengo summarises the dependency theory approach and finds that these theorists

> would agree that at the core of the dependency relation between center and periphery lays [lies] the inability of the periphery to develop an autonomous and dynamic process of technological innovation. Technology – the

[3] K. Smith, "Innovation and growth in resource-based economies", in CEDA, *Competing from Australia* (Australia, 2007), p. 53.

> Promethean force unleashed by the Industrial Revolution – is at the center of stage. The Center countries controlled the technology and the systems for generating technology.[4]

Latin American countries gradually imported fewer manufactured goods in exchange for their exports of natural resource products. This, in turn, led to deficits in trade balances and subsequently economic underdevelopment.[5] In a similar line of argument, the famous paper by the economists Jeffrey D. Sachs and Andrew M. Warner "Natural resource abundance and economic growth" compared data from a wide range of countries and found that economies with natural resources as a large share of exports in 1970 grew slowly during the following two decades.[6] In some cases, large natural resource industries have led to the so-called 'Dutch disease'. In such cases, resource industries have caused a too strong currency for other export products, and have resulted in a decline in manufacturing, or other, industries. The term is related to the decline of the manufacturing sector in the Netherlands after the discovery of natural gas in 1959.[7]

A vast amount of literature finds that natural resources destroy institutions and in some cases create civil wars. Sudan, Nigeria, Angola, and Congo are examples of countries with such problems.[8] The challenge is that while natural resources tend to implicate large incomes fast, they can also be fluctuating. The volatility which these industries represent often destabilises public regimes, weakens state capacity, and encourages rent-seeking behaviour and corruption.[9] In Nigeria, for instance, oil

---

[4] M. Vernengo, "Technology, Finance, and Dependency: Latin American Radical Political Economy in Retrospect", *Review of Radical Political Economics,* Vol. 38, No. 4 (2006), pp. 552–553.

[5] R. Prebisch, *The Economic Development of Latin America and Its Principal Problems* (New York, 1950).

[6] J. D. Sachs and A. M. Warner, "The curse of natural resources", *European Economic Review* 45, 4–6 (2001), pp. 827–838.

[7] "The Dutch Disease" *The Economist,* (1977), pp. 82–83.

[8] P. Collier and A. Hoeffler, "Greed and grievance in civil wars", *Oxford Economic Papers,* 56 (2004), pp. 663–695.

[9] See, for example, T. L. Karl, *The Paradox of Plenty* (Berkeley, 1997); S. Andrade and J. Morales, "The Role of the Natural Resource Curse in Preventing Development in Politically Unstable Countries: Case Studies of Angola and Bolivia", Institute for Advanced Development Studies, Development Research *Working Paper Series* No. 11 (2007); F. van der Ploeg, "Natural Resources: Curse or Blessing?", *Journal of Economic Literature,* (2011).

extraction has changed politics and governance. Military dictatorships have plundered large amounts of wealth gained from this production, which has contributed to miserable economic performance.[10] Political scientist Terry Lynn Karl examines oil-exporting countries in the 1970s and finds that governments in Venezuela, Iran, Nigeria, Algeria, and Indonesia rely too heavily on income from natural resource industries. Public spending increased, but without maintaining a general tax regime and long-term fiscal balance.[11]

Other analyses indicate that natural resources have had negative effects on social structures and human capital formation. According to economists Thorvaldur Gylfason and Gylfi Zoega, dependency on natural resources is often accompanied by greater social inequality.[12] Natural resources have also appeared to weaken the incentives to invest in human capital. This is based on the idea that natural resource industries are founded on medium and low technological activities, which in turn create very few qualified jobs and therefore lower incentives to develop good education systems compared to countries with fewer natural resources. The underdeveloped education systems slow down the pace of economic development.[13] The authors Elena Suslova and Natalya Volchkova test human capital formation and find that industries which require "sophisticated human capital inputs" would be at a disadvantage in natural resource–rich countries.[14]

Thus, in many cases, and in multiple countries, natural resources seemed to have had multiple negative impacts on institutions, social structures, and economic development. Underlying these negative economic effects lies the notion that natural resources generate sectors with medium and low technological activities that are less knowledge-intensive

---

[10] F. van der Ploeg, "Natural Resources: Curse or Blessing?", *Journal of Economic Literature*, (2011), p. 2.

[11] T. L. Karl, *The Paradox of Plenty* (Berkeley, 1997).

[12] T. Gylfason and G. Zoega, "Inequality and Economic Growth: Do Natural Resources Matter?", in Inequality and Growth: Theory and Policy Implications, Theo Eicher and Stephen Turnovsky (eds), (MIT Press, 2003).

[13] T. Gylfason, "Natural Resources, Education and Economic Development", *European Economic Review 45* (2001), pp. 847–859.

[14] E. Suslova and N. Volchkova, "Human Capital, Industrial Growth and Resource Curse", *State University Higher School of Economics: WP 13/2007/11* CAS Working Paper Series, p. 4.

and innovative than manufacturing sectors. Although economists recognise that natural resources provide opportunities for economic growth, it is held that natural resource industries have weak dynamic development patterns, lack linkages with the wider economy, and are less knowledge-intensive and productive than manufacturing or "high-tech" industries.[15]

However, the idea that natural resources are unfavourable to economic growth was questioned early. Economist Jacob Viner declared that "[t]here are no inherent advantages of manufacturing over agriculture" and, in 1955, Douglass North argued against Prebisch's hypothesis, and wrote that "the contention that regions must industrialize in order to continue to grow […] [is] based on some fundamental misconceptions".[16] In this line of argument, comparative analyses by economist Angus Maddison show that resource-rich countries from 1913 to 1950, including Latin American economies, actually grew faster than industrialised countries.[17] Research on specific natural resource industries finds that agriculture and mining industries in some cases experience higher productivity growth than manufacturing industries.[18]

A key empirical problem with the resource curse hypothesis is that some of the richest countries in the world, such as Norway, Sweden, Canada, New Zealand, Australia, and also the United States, have developed highly successful and fast-growing economies based on natural resources. Historically, natural resource industries, notably agriculture, timber, fish, metal, and minerals, have constituted very large parts of these countries' economies. The interesting thing about these cases is that they have developed fast-growing economies instead of

---

[15] For a review of literature on this subject, see B. Nelson and A. Behar, "Natural Resources, Growth and Spatially-Based Development: A View of the Literature", world development report, *Reshaping Economic Geography*, 7 February 2008.

[16] J. Viner, "International Trade and Economic Development", Louvain Economic Review, volume 20, issue 1 (February 1954), p. 72; D. North, "Location Theory and Regional Economic Growth", *Journal of Political Economy* 63 (1955), p. 252.

[17] D. De Ferranti et al. *From Natural Resources to the Knowledge Economy* (Washington, DC, 2002), p. 6.

[18] W. Martin, and D. Mitra, "Productivity Growth and convergence in Agriculture and Manufacturing", Economic Development and Cultural Change, vol. 49, issue 2 (2001), 403–22; G. Wright, G. and J. W. Czelusta, "Resource-Based Growth Past and Present", in D. Lederman and W. F. Maloney (eds.), *Natural Resources: Neither Curse nor Destiny*, chapter 7 (Washington, DC, 2007).

experiencing slow growth, stagnation, and underdevelopment. Keith Smith points out that

> [t]hese small, open economies have rested their development paths on resource-based sectors, and out of them have developed low- and medium-technology industries that have driven growth within these countries. This has been the case not only historically, but in many instances remains the case today.[19]

Moreover, these countries have few social differences; they are open and transparent democracies with a low degree of corruption and rent-seeking. This leads to the hypothesis that natural resource sectors have not obstructed long-run development in these countries.

Norway—the focus of this analysis—has had an abundance of natural resources, mainly waterfalls, forests, fish, minerals, and metallic ores, and natural resource industries have accounted for the largest shares of GDP and exports.[20] Timber was the largest export industry in the late nineteenth century and stood for more than 40 per cent of export in the 1870s. The industry branched out a cellulose and paper industry, and their share of exports remained important, but had reduced their share of exports to 14.2 per cent in 1913. Export of edible animal products (mostly fish) totalled more than 40 per cent in the late nineteenth and early twentieth centuries. In addition to agricultural products, minerals, metals, and chemicals were essential export products, especially from the early twentieth century with the development of new chemical- and electro-metallurgical productions, notably aluminium—made of the aluminium ore bauxite—and artificial fertiliser.[21] Production of aluminium is still today one of the largest in the world. The extraction of oil and gas began in the 1970s and developed to become the most important industry of the economy. Some years it has accounted for more than 50 per

[19] K. Smith, "Innovation and growth in resource-based economies", in CEDA, *Competing from Australia* (Australia, 2007), p. 53.

[20] F. Hodne and O. H. Grytten, *Norsk økonomi i det nittende århundre* (Bergen, 2000), p. 97.

[21] O. Wicken, "The Norwegian Path Creating and Building Enabling Sectors", working paper, Centre for Technology, Innovation and Culture, University of Oslo, 2010, p. 17; Norges Officielle Statistikk, *Nasjonalregnskap 1900–1929* (Oslo, 1953); Norges Offisielle Statistikk, *Nasjonalregnskap 1930–1939 og 1946–1951* (Oslo, 1952).

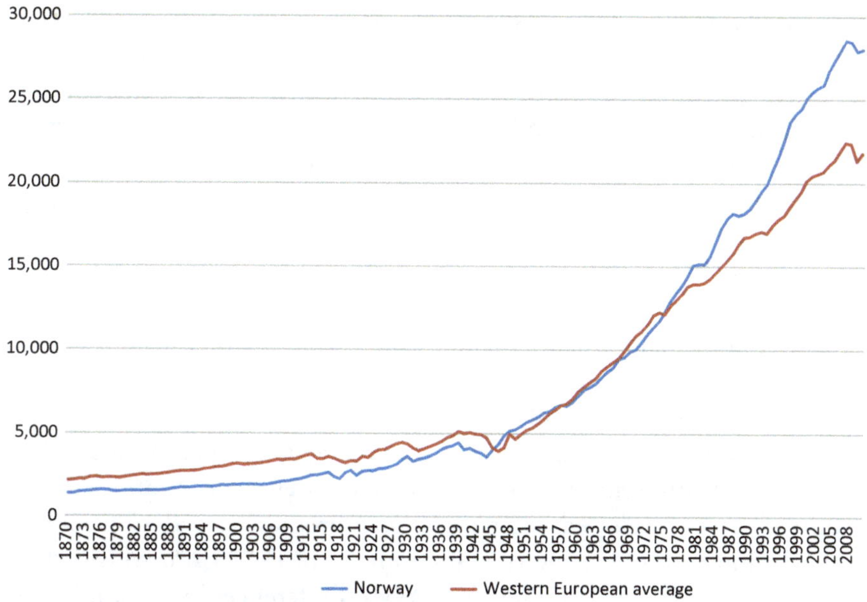

Fig. 1.1 GDP per capita (1990 Int. GK$). (Source: Angus Maddison, *Historical Statistics of the World Economy*: update, 1 January 2013)

cent of exports. In 2012 oil and gas accounted for around 68 per cent of total merchandise exports in Norway.[22]

Norway did not always have the fast-growing economy that it has today. Until the 1930s, Norway was below the Western European average and had relatively slow growth. It was after the Second World War that Norway took a step forward and surpassed the Western European average. Norway is today one of the richest countries in the world and productivity and income are amongst the highest, perhaps even without the extra contribution of the country's oil and gas sector[23] (see Fig. 1.1).

[22] World Data Bank, World Development Indicators (WDI) & Global Development Finance (GDF).

[23] J. Fagerberg et al. "Innovation-systems, path-dependency and policy: The co-evolution of science, technology and innovation policy and industrial structure in a small, resource-based economy", DIME Working paper in the series on "Dynamics of Knowledge Accumulation, Competitiveness, Regional Cohesion and Economic Policies", 2008.

It should also be emphasised that economic growth in Norway has been based on a fairly equal income distribution. Nobility privileges were abolished with independence from Denmark in 1814 and the country had practically no land nobility, Catholic Church, or military caste.[24] Feudal tendencies were weak and by the late nineteenth century, small private farmers owned most of the soil. Self-owned farming represented 81 per cent in 1855 and increased to 95 per cent in 1875.[25] Social differences have been small historically and Norway is today one of the countries in the world with the least social difference.[26] The country has developed and maintained a welfare state, which covers health, education, and social protection for the whole population and is ranked the highest in the Human Development Index.[27]

The key point to be made here is that there are marked differences in economic growth amongst natural resource–intensive economies. Some countries have developed extensively, and progressed and transformed into becoming amongst the richest countries in the world, while others remain underdeveloped and relatively poor. Differences in economic performance across natural resource–intensive economies suggest that an abundance of natural resources does not necessarily lead to stagnation. Conversely, some countries arguably have developed *because* of their natural resources, and not despite them. Economists Magnus Blomström and Patricio Meller compare Latin America and Scandinavia and explain that, despite similar conditions, these two groups of countries have developed very differently:

[24] H. Hveem, "Developing an Open Economy: Norway's Transformation, 1845–1975", in *Diverging Paths Comparing a Century of Scandinavian and Latin American Economic Development,* M. Blomström and P. Meller (eds.), (Washington, DC, 1991), p. 130; F. Hodne and O. H. Grytten, *Norsk økonomi i det nittende århundre* (Bergen, 2000), pp. 16–17.

[25] T. Bergh et al. *Norge fra u-land til i-land: vekst og utviklingslinjer 1830–1980*, (Oslo, 1983), p. 36.

[26] Norway has had small differences in income distribution according to UNDP Income Gini coefficient: Gini Coefficient. Retrieved from UNDP: http://hdr.undp.org/en/content/income-gini-coefficient [accessed 2 March 2017]. It measures the deviation of the distribution of income among individuals or households within a country from a perfectly equal distribution. A value of 0 represents absolute equality; a value of 100, absolute inequality.

[27] United Nations Development Programme, Human Development Reports: http://hdr.undp.org/en/data [accessed 13.04 2018].

> Although the physical characteristics of several Latin American countries have been similar to those in Scandinavia – in both regions, countries have been small in area and population and rich in natural resources – Latin America has developed very differently. While the northern countries were able to use their natural resources to generate the momentum for sustained growth, the countries of Latin America have tried different formulas without much success.[28]

Large differences between natural resource–intensive economies suggest that some of the research on natural resource–intensive economies is misleading. A severe problem in much of the literature which finds that natural resources inhibit growth is that these conclusions are too general and that differences in growth are not accounted for or explained. Based on these observations, Smith, amongst others, criticises the generalisation that has been made to claim slow growth amongst natural resource–intensive economies. He stresses that

> if the data on resource-based economies [are not] as secure as it might be, then it may be that the problems that the "resource curse" hypothesis is seeking to explain are not as general as they seem to be. This leaves us with an interesting question: what factors explain growth in successful resource-based economies?[29]

I find that it is essential to complement the literature with more systematic and in-depth empirical analyses. Only this way can we further our understanding of how economies based on natural resources develop and grow. The aim of this investigation is to explore in more detail some of the underlying factors that may explain how and why natural resource–intensive economies grow by investigating a natural resource industry with long traditions in many countries, namely mining, and I explore this by looking at Norway, one of the richest natural resource–intensive economies in the world.

---

[28] M. Blomström and P. Meller (eds.), *Diverging Paths Comparing a Century of Scandinavian and Latin American Economic Development* (Washington, DC, 1991), p. vii.

[29] K. Smith, "Innovation and growth in resource-based economies", in CEDA, *Competing from Australia* (Australia, 2007), p. 7.

Natural Resources and Knowledge Development

Why, then, have some natural resource–intensive economies experienced strong economic growth, while others have not? What should we focus on? Which factors determine growth in successful natural resource–intensive countries? Some scholars have recently started to address these questions. Instead of assuming that natural resource industries result in poor economic performance, scholars now focus on factors which actually determine growth in natural resource sectors and natural resource–intensive countries.

First and foremost, there is the issue of how much of the natural resources is utilised. New analyses suggest that there have been large differences when it comes to the utilisation of the natural resource potential. Fundamental to the new ways of approaching natural resources and growth is that natural resources are not geologically predetermined, but endogenous and socially constructed. Natural resources, and the extraction of these, are not given, but needs and usages of natural resources have determined their extraction and transformation. Cali Nuur and Staffan Laestadius demonstrate that the development of natural resource industries and their application areas have changed over time:

> Charcoal was important for steel production until [the] mid nineteenth century but its role diminished rapidly after coal became the dominant source of energy for industrial processes. Later the large waterfalls became a Scandinavian energy resource compensating for lack of coal. And the phosphorous iron ores in Sweden increased dramatically in value with the Gilchrist-Thomas process. Radical technical change – which does not fall like manna from heaven but is the result of capabilities created for some reason somewhere – may thus cause tough incentives to industrial transformation as knowledge related to a specific, and maybe local, natural resource becomes obsolete.[30]

[30] C. Nuur and S. Laestadius, "Anomalies in the resource curse paradigm: the case of Sweden". Trieta research report, The Royal Institute of Technology, KTH, 2009, pp. 46–47.

Countries have differed greatly when it comes to utilisation of initial stocks of natural resources. In their historical analysis of mineral and metal extraction, economists Paul David and Gavin Wright find that the mining sector in the United States benefitted from their mineral resources to a far greater extent than any other country between 1850 and 1950, even though other countries initially had more mineral reserves. Mineral-rich countries, such as Brazil, Chile, Russia, and Australia, had a much smaller production than the mineral deposits would suggest. Chile—a country with one of the largest copper deposits in the world—had large unexploited mineral and metal deposits and the country's copper production was far below its proportionate share of the world's copper resources.[31] Beyond this, there are questions about diversification, and about how resource industries can create linkages to other industries. Mining companies, especially large-scale mining corporations, have depended on energy supply and input of machinery, equipment, engineering, infrastructure, technical services, and finances, and such firms have created important backward linkages.[32] Economist David Ferranti and colleagues suggest that

> [t]he fact that Australia, Canada, and the Scandinavian countries succeeded by playing to their resource strengths, suggests that success has less to do with what a country produces in particular, and everything to do with the way in which it produces it.[33]

Perhaps the most severe problem with the literature arguing for the resource curse is the assumption that natural resource industries are based on simple knowledge and working operations, which only require unskilled workers. In the case of mining—a typical natural resource industry—would not the mapping of ore deposits and analyses of minerals and metals require knowledge specialisations and workers with in-depth knowledge in geology, mineralogy, physics, mechanics, and chemistry?

---

[31] P. David, G. and Wright, "Increasing Returns and the Genesis of American Resource Abundance", *Oxford Journals*, vol. 6, issue 2 (1997), pp. 210–218.

[32] UNCTAD, *World Investment Report 2001 Promoting Linkages* (New York and Geneva, 2001).

[33] D. De Ferranti et al. *From Natural Resources to the Knowledge Economy* (Washington, DC, 2002), p. 52.

Do the planning and making of mines not entail some of the most complex constructions that exist? After all, they are normally underground and involve tunnels, adits, and ventilation systems, which demand precise mathematical calculations and scientific principles to prevent from collapsing. Conversely, evidence suggests that natural resource–intensive industries in high-income economies have been highly knowledge-intensive, dynamic, and innovative; they have created linkages to other industries within the economy; and they have developed specialisations and new industries which have contributed to complex economic structures.[34]

More specifically, differences in the way natural resources are produced, and how natural resource industries develop, suggest varieties in knowledge foundations. Historian Olav Wicken stresses that different knowledge bases can encourage knowledge specialisations and develop competitive firms and industries, which in turn can lead to dissimilar trajectories between such economies.[35] Analyses of specific natural resource sectors support this theory. In a recent study Hartmut Hirsch-Kreinsen, David Jacobsen, Staffan Laestadius, and Keith Smith showed that current fish production, in some countries, is based on multiple knowledge systems:

> Examples of embodied flows in fishing include use of new materials and design concepts in ships, satellite communications, global positioning systems, safety systems, sonar technologies (linked to winch, trawl and ship management systems), optical technologies for sorting fish, computer systems for real-time monitoring and weighing of catches, and so on. Within fish farming, these high-technology inputs include pond technologies (based on advanced materials and incorporating complex design knowl-

[34] See, for example, A. D. Andersen, "Towards a new approach to natural resources and development: the role of learning, innovation and linkage dynamics". *Int. J. Technological Learning, Innovation and Development*, vol. 5 2012. (3); D. De Ferranti et al. *From Natural Resources to the Knowledge Economy* (Washington, DC, 2002); H. Hirsch-Kreinsen et al., "Low-Tech Industries and the Knowledge Economy: State of the Art and Research Challenges", *PILOT Policy and Innovation in Low-Tech, STEP – Centre for Innovation Research* (Oslo, 2003); S. Ville and O. Wicken, "The Dynamics of Resource-Based Economic Development: Evidence from Australia and Norway", *Industrial and Corporate Change*, Volume 22, Issue 5, (1 October 2013).

[35] O. Wicken, "The Norwegian Path Creating and Building Enabling Sectors", working paper, Centre for Technology, Innovation and Culture, University of Oslo, 2010, pp. 4–5.

> edges), computer imaging and pattern recognition technologies for monitoring (including 3D measurement systems), nutrition technologies (often based on biotechnology and genetic research), sonars, robotics (in feeding systems), and so on.[36]

Wicken explains differences in knowledge bases and specialisations further. He stresses that natural resource–intensive industries and economies should be examined

> as dynamic processes which can evolve in different directions depending on social and political contexts as well as by decisions undertaken by organizations and individuals. Economies specializing in natural resources may therefore follow different historical paths of economic and social evolution.[37]

He, and economic historian Simon Ville, compare two successful natural resource–intensive economies historically, Norway and Australia, and find that linkages have developed between natural resource–based industries and other sectors, such as capital goods sectors, services, and research institutions. New industries based on natural resources have often emerged out of these linkages and been vital for growth in both countries. They conclude that "dynamic interactive relationship between natural resource industries and enabling sectors is regarded as the core aspect of the successful economic development of Australia and Norway".[38] The development of the high-performing oil and gas industry in Norway from the 1970s is well known. Studies show that state organisations, oil companies, universities, and research centres collaborated and interacted with each other, which facilitated capacity building and the spread of relevant knowledge. The Norwegian government played a

---

[36] H. Hirsch-Kreinsen et al., "Low-Tech Industries and the Knowledge Economy: State of the Art and Research Challenges", *PILOT Policy and Innovation in Low-Tech, STEP – Centre for Innovation Research* (Oslo, 2003), p. 22.

[37] Wicken, "The Norwegian Path Creating and Building Enabling Sectors", working paper, Centre for Technology, Innovation and Culture, University of Oslo, 2010, p. 2.

[38] S. Ville and O. Wicken, "The Dynamics of Resource-Based Economic Development: Evidence from Australia and Norway", *Industrial and Corporate Change*, Volume 22, Issue 5, (1 October 2013), p. 37.

role in negotiating with—and regulating—domestic and foreign oil companies, which contributed to the development of key research.[39] History is used to identify the development of the United Kingdom and Germany—two coal-abundant economies during the Industrial Revolution—by looking at how they took advantage of their natural resources.[40] Paul David and Gavin Wright analyse the underlying institutional foundations for the successful mining sector in the United States and show that new knowledge and techniques enabled growth in mining as well as connections to other parts of the economy, including other non-natural resource sectors:

> We find […] that late nineteenth century American mineral expansion embodied many of the features that typify modern knowledge-based economies: positive feedbacks to investments in knowledge, spillover benefits from one mining specialty to another, complementarities between public- and private-sector discoveries, and increasing returns to scale—both to firms and to the country as a whole.[41]

Some natural resource–based economies have managed to benefit from their natural resources, develop competitive natural resource sectors, and have strong economic growth, while others have stagnated and declined. Instead of assuming that natural resources lead to slow growth, the notion is that industries are dependent on society and an encouraging institutional environment to make the most of natural resources. Considering the previous, how have innovative, dynamic, knowledge-based, productive, and high-performing natural resource industries developed? What drives progress in such industries? The next section discusses some of the general literature about institutions and economic growth and sets the basis for the

[39] B. Sæther, A. Isaksen and A. Karlsen, "Innovation by co-evolution in natural resource industries: The Norwegian experience", *Geoforum*, vol. 42, Issue 3, June (2011); Å. Cappelen and L. Mjøset, "Can Norway be a Role Model for Natural Resource Abundant Countries?", in *Development Success: Historical Accounts from More Advanced Countries*, A. K. Fosu (ed.) (Oxford, 2013); E. R. Larsen, "Are Rich Countries Immune to the resource Curse? Evidence from Norway's Management of Its Oil Riches", Discussion paper 362 Statistics Norway (Oslo, 2003); J. A. Frankel, "The Natural Resource Curse: A survey", NBER Working paper 15836 (Cambridge, 2010).

[40] Wrigley, E. A. *Energy and the Industrial Revolution*. Cambridge (Massachusetts, 2010).

[41] P. David, G. and Wright, "Increasing Returns and the Genesis of American Resource Abundance", *Oxford Journals*, vol. 6, issue 2 (1997), pp. 204–205.

analysis of certain "knowledge organisations" which seemed to be directly involved in accumulating and developing knowledge for mining in Norway.

Institutions, Incentives, and Economic Growth

What is the foundation of modern economic growth, and why is there such inter-country variation within it? Why are some countries rich and others poor? More specifically, how can we analyse development of natural resource–intensive industries? A strong argument in economic history is that knowledge is the underlying basis of economic growth and that a substantial increase in "useful knowledge" is the basis for the economic development that has occurred since the late eighteenth century. In 1965, Nobel Laureate in economics Simon Kuznets wrote that useful knowledge is the source of modern economic growth.[42] The Nobel Laureate in economics Douglass C. North stated later that "modern economic growth had as its source the growth in the stock of knowledge".[43] Kristine Bruland and economist Keith Smith confirm that

> however much growth may require capital accumulation and changes in the quantity and quality of labour, its ultimate source is technological capability. Capital and labour can only be deployed around specific technologies, and the capability to use them. Such capability in turn is a function of learning, of knowledge accumulation.[44]

Joel Mokyr further explains that the enormous increase in the knowledge base that has occurred during the past two centuries is the foundation of modern economic growth and the material world as we know it today. He finds that throughout history, the lack of knowledge has been the principal reason why societies have been limited in their ability to

[42] S. Kuznets, *Economic Growth and structure* (New York, 1965), p. 84.

[43] D. C. North, "Economic Performance Through Time: The Limits to Knowledge", EconWPA, Economic History, 9612004 (1996), p. 9.

[44] K. Bruland and K. Smith, "Knowledge Flows and Catching-Up Industrialization in the Nordic Countries The Roles of Patent Systems", in H. Odagiri et al. *Intellectual property rights, development and catch-up an International comparative study* (Oxford, 2010), pp. 63–94.

increase material wealth. In his analysis of why the Industrial Revolution happened first in Britain, and not somewhere else, Mokyr shows that the West acquired the necessary useful knowledge to undergo a sustained economic progress which had not existed before. Inventions and technological changes had occurred before industrialisation, but the new at this point in history was that innovation processes continued, and acceleration of technological changes did not stop or fade out.[45] In the same line of argument Kline and Rosenberg point out that "[t]he relevant questions are not whether innovation is necessary to increases in efficiency or for survival, but rather: what kind of innovations? At what speed?"[46]

In this context, there is a general agreement that learning and innovation depend on institutional support. The capacity and decision to innovate are intimately linked to the context within which they occur. Bengt-Åke Lundvall confirms that "learning is predominately an interactive and, therefore, a socially embedded process which cannot be understood without taking into consideration its institutional and cultural context".[47] Perhaps the most influential economist in this respect is Douglass C. North. He understood "institutions" as "the rules of the game", including formal laws, regulations, informal norms, and values, which regulate economic behaviour. "Organisations" are players, actors, or groups of individuals bound by a common purpose to achieve objectives. Institutions and organisations affect, influence, and interact with each other.[48] Charles Edquist uses North's definitions to explain the relations between institutions and organisations in the following way:

> The relations between organisations and institutions are important for innovations and for the operation of systems of innovation. Organisations are strongly influenced and shaped by institutions; organisations can be said to be "embedded" in an institutional environment or set of rules, which include the legal system, norms, standards, etc. But institutions are also

---

[45] J. Mokyr, *The Enlightened Economy* (New Haven, 2009), pp. 4, 17, and 291.

[46] D. Kline and N. Rosenberg, "An Overview of Innovation", in R. Landau and N. Rosenberg (eds.) *The Positive Sum Strategy: Harnessing Technology for Economic Growth* (Washington, DC, 1986), p. 279.

[47] B.-Å. Lundvall, *National Systems of Innovation* (London, 1992), p. 1.

[48] D. C. North, "New institutional economics and development" (Washington, DC, 1993), pp. 6 and 97.

> "embedded" in organisations. Examples are firm specific practices with regard to bookkeeping or concerning the relations between managers and employees; a lot of institutions develop inside firms. Hence, there is a complicated two-way relationship of mutual embeddedness between institutions and organisations, and this relationship influences innovation processes and thereby also both the performance and change of systems of innovation.[49]

Economists and historians search for differences in institutions and organisations when explaining variances in growth amongst countries. Simon Kuznets affirmed in 1965 that "the capacity of the society to devise and accept the institutional changes that may be necessary for [...] changes in technology and substance of economic production [is indispensable]".[50] North found that an institutional framework, which promotes, or at least does not hinder, technological change and learning is a requirement for innovation.[51] Kristine Bruland follows this approach and sums up the idea: "[W]e might expect to find significant inter-country differences in how learning occurs, in terms of institutions and organisations, and this is likely to have important implications for histories of economic growth."[52]

However, scholars focus on different institutions and organisations in their analyses. There are many types of organisations and institutions, they function in different ways, and they might affect learning and innovation differently. A recurring approach is to examine institutions and organisations in relation to whether they constrain or act as an incentive to economic transactions. In this line of thought, scholars tend to characterise them according to their "quality" and classify them as either "good" or "bad" or "efficient" or "inefficient". Many follow North's argument, which is that there are lower transaction costs when institutions are efficient, and the costs are higher when institutions are inefficient, or a

---

[49] C. Edquist, "The Systems of Innovation Approach and Innovation Policy: An account of the state of the art" (Aalborg, 2001), p. 6.

[50] S. Kuznets, *Economic Growth and structure* (New York, 1965), pp. 111–112.

[51] D. C. North, "Institutions" in *The Journal of Economic Perspectives*, Vol. 5, No. 1, (1991), p. 97.

[52] K. Bruland, "Skills, learning and the International Diffusion of Technology: a Perspective on Scandinavian Industrialization", in M. Berg and K. Bruland (eds.), *Technological Revolutions in Europe* (Cheltenham, 1998), p. 161.

hindrance to economic growth.[53] Depending on the setting, institutions enable possibilities, or they set limits to human behaviour. Economist Avner Greif affirms that good institutions

> provide the foundations of markets by efficiently assigning, protecting, and altering property rights; securing contracts; and motivating specialization and exchange. [They] also encourage production by fostering saving, investment in human and physical capital, and development and adoption of useful knowledge. They maintain a sustainable rate of population growth and foster welfare-enhancing peace; the joint mobilization of resources; and beneficial policies, such as the provision of public goods.[54]

An important point here is that institutions are found to develop slowly and to persist over time. Douglass North explains the "path dependence" of institutions in the following way:

> [H]istory [...] is largely a story of institutional evolution in which the historical performance of economies can only be understood as a part of a sequential story. Institutions provide the incentive structure of an economy; as that structure evolves, it shapes the direction of economic change towards growth, stagnation, or decline.[55]

But which *types* of institutions, and organisations, are found to influence learning and innovation? Alexander Gerschenkron wrote in 1962 that Britain industrialised first because it was early in developing certain institutional and technological conditions, which he regarded as necessary "prerequisites" for industrialisation to take place. These included entrepreneurship, an industry-oriented financial system, and an effective labour market. When other European countries industrialised, they could do so because they had succeeded in developing efficient "substitutes" for any prerequisites that they missed. According to Gerschenkron, "certain major obstacles to industrialization must be removed", such as feudal

---

[53] D. C. North, "Institutions" in *The Journal of Economic Perspectives*, Vol. 5, No. 1, (1991), p. 97.

[54] A. Greif, *Institutions and the Path to the Modern Economy: Lessons from Medieval Trade* (Cambridge, 2006), p. 4.

[55] D. C. North, "Institutions" in *The Journal of Economic Perspectives*, vol. 5, No. 1 (1991), p. 97.

obstructions to movement of people and goods, and "certain things propitious to it must be created before industrialization can begin".[56] These "propitious things" are the prerequisites referred to earlier. Scholars have continued to develop models of industrialisation and to focus on preconditions or "prerequisites" for economic growth. In general terms, democracy, freedom, openness, transparency, and honesty are found to reduce corruption and rent-seeking; education and training build human capital and a more productive workforce; secure property rights motivate savings and investments; an organised system for imports and exports facilitates trade and economic stability and reduces inflation and volatility. In cases in which institutions do not favour development of industries, slow growth or stagnation could follow.[57]

The state clearly influences economic activities in different ways. Public institutions regulate taxes, property rights, the monetary system, and imports and exports through tariffs; they finance education and research, infrastructure, and other national projects; and they may redistribute wealth and control the utilisation of natural resources through regulations and provision of concessions. However, there is no consensus with regard to *how* public institutions would best relate to the economy. Should the state regulate markets and protect domestic industries through tariffs on imports, or should it not interfere in the economy and assure free trade? Economists have a long tradition of discussing "right" market incentives and institutional constraints and they disagree on whether free trade or protectionism is better to achieve material prosperity and growth.[58] Throughout the nineteenth century, scholars such as David Ricardo, Richard Cobden, and John Bright were in favour of free trade. Later, politician Alexander Hamilton and economist Friedrich List stated that free trade was unfavourable trade policy for new nations that had not developed internationally competitive industries. They defended taxes on imported finished goods as part of the fostering of national industries.[59]

---

[56] A. Gerschenkron, *Economic backwardness* (Cambridge, 1962), p. 31.

[57] See, for example, A. Greif, *Institutions and the Path to the Modern Economy: Lessons from Medieval Trade* (Cambridge, 2006); D. North et al., *Violence and social orders* (Cambridge, 2009); D. Acemoglu and J. A. Robinson, *Why Nations Fail* (London, 2012).

[58] F. Sejersted, *Demokratisk kapitalisme* (Oslo, 1993), p. 46.

[59] F. Hodne and O. H. Grytten, *Norsk økonomi i det nittende århundre* (Bergen, 2000), p. 13.

There are heavy disagreements related to what role the state should play in the economy, but there is a stronger consensus that democracies allow for more efficient utilisation of natural resources and encourage innovation to a larger degree than authoritarian regimes. The theory is that a broader part of the population has access to property rights, wealth and education and actors are free to organise, operate, and trade in democratic societies. Within the resource curse debate, the main argument is that political institutions in "successful" natural resource–intensive economies facilitate incentives to make profits of natural resources, while they do not in less successful ones. Malebogo Bakwena, for example, finds that democratic regimes generate more growth from natural resources than non-democratic regimes.[60] Similarly, economists Rabah Arezki and Markus Bruckner focus on the twentieth century and show that incomes from natural resources in democracies increased GDP per capita, while GDP per capita decreased in autocracies.[61] Economists Halvor Mehlum, Karl Moene, and Ragnar Torvik distinguish between "grabber-friendly" and "producer-friendly" institutions and look at rule of law, bureaucratic quality, corruption in government, risk of expropriation, and risk of government repudiation of contracts. They make broad categories of natural resource–intensive economies and find that grabber-friendly institutions reduce income, while producer-friendly institutions increase income.[62] Much in the same way, Sambit Bhattacharyya and Roland Hodler study corruption and resource wealth, and predict that the rents from natural resources lead to an increase in corruption if the quality of institutions is poor, while this is not the case otherwise.[63]

[60] M. Bakwena, "Do Better Institutions Alleviate the Resource Curse? Evidence from a Dynamic Panel Approach", PhD Thesis, The University of Queensland Publication (April 2010). See also M. Bakwena, P. Bodman, T. Le and T. Ki, "Avoiding the Resource Curse: The Role of Institutions" University of Queensland, Australia, No 32, (2009) MRG Discussion Paper Series from School of Economics, University of Queensland, Australia; A. Cabrales and E. Hauk. "The quality of political institutions and the curse of natural resources", *The Economic Journal*, 121 (February 2009).

[61] R. Arezki, R. and M. Bruckner, "Resource Windfalls and Emerging Market Sovereign Bond Spreads: The Role of Political Institutions", University of Adelaide, School of Economics, *Research Paper* No. 2011-08 (January 2011).

[62] H. Mehlum, K. Moene and R. Torvik, "Cursed by resources or institutions?" (Oslo, 2005), p. 1.

[63] Bhattacharyya, Sambit and Roland Hodler. "Natural Resources, Democracy and Corruption". *OxCarre Research Paper*, No. 2009-20 (2009).

Political institutions are also used extensively in general models of development and economic growth. Acemoglu and Robinson study several world regions and find that what matters to economic growth is political institutions, which in turn determine economic institutions.[64] They present broad historical comparisons in which countries are deliberately classified into two main categories and their model is based on what the authors define as "extractive" and "inclusive" political and economic institutions. Inclusive political institutions are institutions which "distribute political power widely in a pluralistic manner and are able to achieve some amount of political centralization so as to establish law and order, the foundation of secure property rights and an inclusive market economy". Inclusive economic institutions are institutions which "enforce property rights, create a level playing field, and encourage investments in new technologies and skills". In such societies, a broad spectrum of the population is included in democratic governing processes and exploitation of people is absent or reduced. Extractive political institutions, for their part, "concentrate power in the hands of a few who will then have incentives to maintain and develop extractive economic institutions for their benefit". Extractive economic institutions are "structured to extract resources from the many by the few and [...] fail to protect property rights to provide incentives for economic activity". In these societies only a small elite have been in control and have encouraged exploitation of the rest of the population. Drawing on historical comparisons of continents, regions, and countries, they find that throughout history inclusive political systems have provided incentives for people to acquire skills, save capital, invest, work hard, and innovate, whereas extractive institutions, in the long run, have not.[65] A problem with such broad models is that they will usually ignore specificities of individual countries and economic sectors. Although all modelling generalises, there is often a serious absence of specific details in accounts of industrialisation. Acemoglu and Robinson develop a model based on relatively few historical cases, in which the complexity of different countries' developments is barely revealed. Moreover, such models tend to ignore the fact that countries have in fact

[64] D. Acemoglu and J. A. Robinson, *Why Nations Fail* (London, 2012), p. 43.

[65] D. Acemoglu and J. A. Robinson, *Why Nations Fail* (London, 2012), pp. 429–430.

followed very different paths to growth.[66] The problem here is that the industrial and economic characteristics of the larger nations have differed considerably from those of smaller countries, such as Benelux, the Nordic countries, Canada, and New Zealand, although all of them have experienced strong economic growth.

Property rights are given a large role in the literature about institutions and economic growth. It is found that organisations will save, invest, and innovate only in cases in which the rights to their property are protected, and without secure property rights, the motives to save, invest, and innovate will be reduced.[67] An efficient legal system and laws on how individuals can control, benefit from, and transfer land property are therefore understood as a very important incentive for economic growth, and secure property rights are regarded as a determinant of economic growth. The same goes for natural resource–intensive economies. According to David and Wright, clear property rights and tax exemptions on mining were essential in the successful development of the mining sector in the United States.[68] Authors William Culver and Cornel Reinhard go further and compare the role of mining laws and land rights in the copper industries in the United States and Chile. They find that efficient mining laws in the United States led to the success of this country's copper industry, while inefficient mining laws explain the declining copper industry in Chile in the late nineteenth century. The Chilean laws were inconvenient and resulted in higher transaction costs than in the United States.[69] Then again, although property rights probably are a prerequisite for progress, it is less clear how clear property rights have led to innovation. Secure property rights may "provide incentives for economic activity", but they do

---

[66] A. Milward and S. B. Saul's work is a rare exception to this. See *The Economic Development of Continental Europe, 1780–1870* (London, 1973) and *The Development of the Economies of Continental Europe* (London, 1977).

[67] For the importance of secure property rights in a historical perspective, see D. North et al., *Violence and social orders* (Cambridge, 2009) and A. Greif, *Institutions and the Path to the Modern Economy: Lessons from Medieval Trade* (Cambridge, 2006).

[68] P. David, G. and Wright, "Increasing Returns and the Genesis of American Resource Abundance", *Oxford Journals*, vol. 6, issue 2 (1997), pp. 210–218.

[69] W. W. Culver and C. J. Reinhart "Capital dreams: Chile's Response to Nineteenth-Century World Copper Competition", *Comparative Studies in Society and History*, vol. 31, No. 4 (October 1989), p. 739.

not explain innovation and economic growth alone. Kristine Bruland highlights this point. She writes that it is often believed that if a set of conditions are in place, innovation and growth will automatically follow.[70] However, accounting for incentives does not explain *how* innovation actually occurs and which actors are involved, and it does not explain *why*, for example, certain actors and companies are successful, while others are not.

Much in the same way, intellectual property rights are found to encourage innovation by granting inventors monopoly of their inventions.[71] The economic historian Bjørn Basberg examines the patent system in Norway between 1860 and 1914 and finds that an increase in patents coincided with industrialisation of the country. The theory here is that the patent system motivated and stimulated the industrialisation process.[72] The traditional argument is that technological change depends on incentives for makers to benefit from their inventions. Yet, other scholars question whether such rights determine innovation. Joel Mokyr, for example, gives the patent system less priority in his explanation of the Industrial Revolution. He finds that other institutions were equally—or more—important for continuous technological change. Not only that, patents were sometimes used strategically to block research in a specific direction, which consequently slowed down innovation. A number of inventors of the Industrial Revolution did not use the patent system at all.[73] Similarly, Bruland and Smith show that other institutions explain innovation to a larger degree than patent systems in the Nordic countries.[74] Studies in recent decades also downgrade the importance of intellectual property rights. A study from 2012 actually suggests that weak intellectual property rights has stimulated research and development

---

[70] K. Bruland, *British technology and European industrialization* (Cambridge, 1989), p. 15.

[71] R. Falvey and N. Foster, "The Role of Intellectual Property Rights in Technology Transfer and Economic Growth: Theory and Evidence" (Vienna, 2006), p. vii.

[72] B. L. Basberg, "Patenting and Early Industrialization in Norway, 1860–1914. Was there a Linkage?", *Scandinavian Economic History Review*, vol. 54, Issue 1 (2006), p. 14.

[73] J. Mokyr, "Intellectual property rights, the Industrial Revolution, and the Beginnings of Modern Economic Growth", *American Economic Review*, vol. 99, issue 2 (2009), p. 351.

[74] K. Bruland and K. Smith, "Knowledge Flows and Catching-Up Industrialization in the Nordic Countries: The Roles of Patent Systems", in H. Odagiri et al. *Intellectual property rights, development and catch-up an International comparative study* (Oxford, 2010), pp. 63–94.

(R & D) activities and encouraged knowledge spillovers.[75] There is no census with regard to patent systems, which makes their role in the innovation processes uncertain. Furthermore, even if clear patent systems motivate inventors to be creative, they do not reveal anything about the process of creating and accumulating knowledge, and *how* technological capabilities develop.

Despite the little focus on the actual innovation processes it is generally affirmed that innovation is the source of modern economic growth. Acemoglu and Robinson confirm that "sustained economic growth requires innovation, and innovation cannot be decoupled from creative destruction, which replaces the old with the new in the economic realm".[76] Much in the same way, North confirmed that "open access societies" have developed because they have supported innovation, and given the freedom to create, through competition:

> Open access societies [...] are competitive. Competition dominates the way in which both political and economic markets work. The economy works by innovative creation; there is competition in markets, and those players who create more efficient, productive methods stand to gain and replace those who are less efficient. Innovation and creativity are the heart of what makes markets work and what has made the modern world so dynamic and such an extraordinary place.[77]

Central to North's hypothesis is that economic growth is based on the ability to support complex, sophisticated organisations. He classifies developed societies according to their open access to organisations and support of organisational complexity. Developed societies are

> filled with a rich variety of complicated and sophisticated organizations capable of producing goods and services, carrying out research and devel-

[75] A. Breitwieser and N. Foster, "Intellectual property rights, innovation and technology transfer: a survey", (Vienna, 2012), p. 2.

[76] D. Acemoglu and J. A. Robinson, *Why Nations Fail* (London, 2012), p. 430.

[77] D. North, "Violence and social orders", The annual proceedings of the Wealth and Well-Being of Nations (2009), p. 22.

opment, and coordinating individual behaviour on a scale never before seen in human history.[78]

Following this argument, the drive which directly influences and supports the development of complex and coordinated organisations should be explored in further detail. An issue with the general theories of institutions and organisations outlined here is that they provide little information about the channels through which knowledge is developed and innovation is generated. They do not detect the direct link between specific organisations and institutions on the one hand and innovation processes on the other. Little empirical information is provided about knowledge accumulation and the processes behind technological change, which in turn characterise successful industries and economies, and the underlying reasons for the less innovative and unsuccessful countries and sectors.

An approach which seems to focus more directly on innovation, and the organisations and institutions involved, is the "National Innovation System approach". With this method, scholars conceptualise and explore networks and interactions based on a given institutional and organisational framework. Economic policy, the role of the state, and investments in R & D are often included in this.[79] Similarly, "Regional Innovation System" and "Sectoral Innovation System" are used to explore such issues only within a smaller framework. The core idea is that institutional and organisational frameworks have varied considerably between countries, regions, and sectors, and this, in turn, has resulted in different kinds of systems.[80]

These approaches focus more on innovation processes, but they are criticised for their heavy emphasis on R & D from a national perspective, which makes such models highly simple.[81] Formal scientific centres and

[78] D. North, "A conceptual framework for interpreting recorded human history", George Mason University, *working paper* 75 (Cambridge, 2006), pp. 3–4.

[79] C. Edquist, "The Systems of Innovation Approach and Innovation Policy: An account of the state of the art" (Aalborg, 2001), p. 2.

[80] See R. R. Nelson, R. R. *National Innovation Systems A Comparative Analysis* (New York, 1993).

[81] C. Edquist, "The Systems of Innovation Approach and Innovation Policy: An account of the state of the art" (Aalborg, 2001), p. 3.

company research laboratories are not the only places where learning and innovation are formed and may not capture the complexity of innovation processes. Natural resource industries, for example, are found to be based less on scientific research and R & D, and more on "learning by doing".[82] A type of organisation, which is hardly included in National Innovation System models, is education and individual learning. Although there is a broad consensus that schooling is important to generate innovation, it is normally left out. Charles Edquist criticises innovation systems approaches and emphasises that they do not grasp or account for the entire complex, random, and disorganised processes of learning and innovation. They may detect some important organisations and institutions but fail to capture the broader set of knowledge accumulation, learning, and innovation progressions. He concludes that

> we know far too little about the determinants of innovation, although this is a weakness of innovation studies in general […] [the systems] approach partly neglects other types of learning process than those leading to innovations in a direct and immediate way.[83]

Looking beyond the National Innovation System approach, there is a wide agreement that basic education has encouraged learning and facilitated an overall skilled workforce. Numerous studies suggest that education is crucial for industrial development and is used extensively in explanations for why some countries have developed and become rich and others have not. Nobel Laureate in economics Gary S. Becker saw schooling, along with on-the-job training and so on, as an investment in human capital and understood education as a factor which directly increases productivity. He made a division between "specific" and "general" human capital. Specific human capital refers to skills that are useful only to a single firm, whereas general human capital, such as literacy, numeracy, knowledge about mechanics and chemistry, are useful in many

---

[82] H. Hirsch-Kreinsen et al., "Low-Tech Industries and the Knowledge Economy: State of the Art and Research Challenges", *PILOT Policy and Innovation in Low-Tech, STEP – Centre for Innovation Research* (Oslo, 2003), p. 9.

[83] C. Edquist, "The Systems of Innovation Approach and Innovation Policy: An account of the state of the art" (Aalborg, 2001) p. 3.

industries.[84] From a historical perspective, whole sets of training systems are found to have been essential to industrial performance:

> At least from the time of the Industrial Revolution, technologies (and their underlying knowledge bases) have become complex enough to require either high standards of literacy and numeracy or more specific training in relevant disciplines (i.e. basic sciences or technological disciplines such as electrical or chemical engineering) if they are to be operated successfully. From this perspective, education is seen as a key input to the advanced economy. More specifically, education is the investment process through which "human capital" is created, and this is increasingly seen as the central input to growth.[85]

High-quality education and infrastructure system are pointed out especially to have facilitated dynamic and knowledge-based natural resource industries. Nicolai Petrovsky finds that education, amongst other factors, may affect natural resource development positively and reduce potential negative economic effects.[86] Mike Smart finds a causal chain from diversity of natural resources and educational attainment to democratisation and from democratisation to per capita GDP.[87] Recent studies come to similar conclusions. Brooks and Kurtz suggest that human capital resources condition development in oil-rich countries, because they "make possible the management of resources in ways that encourage the absorption of technology and development of new economic sectors". When human capital is absent, however, the resource curse is more likely to present itself.[88] Ronald Mendoza, Harold McArthur, and Anne Ong Lopez argue that "notably human capital – appear[s] to

[84] G. S. Becker, *Human Capital* (New York, 1975), pp. 19–37.

[85] K. Bruland, "Education" in *Oxford encyclopedia of economic history, vol. II* (Oxford, 2003), pp. 161–162.

[86] N. Petrovsky, "Does Natural Resource Wealth Spoil and Corrupt Governments? A New Test of the Resource Curse Thesis", thesis, August 2004; Denton, Texas.

[87] M. Smart, "Natural Resource Diversity and Democracy", *The Economic Society of Australia*, vol. 28 (December 2009), pp. 366–375.

[88] S. M. Brooks and M. Kurtz, "Conditioning the "Resource Curse": Globalization, Human Capital, and Growth in Oil-Rich Nations", *Comparative Political Studies,* 44(6), (March 2011).

have transformed the natural resource curse into a boon for development".[89] Gøril Bjerkhol Havro and Javier Santiso draw on experiences from Chile and Norway and show that both countries have benefited from natural resources in recent years. They suggest that natural resource wealth is dependent on macro-economic policy, but also on capable civil servants, a developed business community, and human capital.[90]

Technical education is also understood as key to technological change and industrial development. During the nineteenth century, many technical educational institutes were established across Europe with the aim of providing technicians, engineers, and other specialists to industries, and there was an increasing conviction that technical education led to industrial development. Peter Lundgreen finds that

> [d]uring the late nineteenth century a general belief in education as a key to international competition joined forces with a thorough transformation of the economy in providing new job markets for engineers graduating from colleges and universities.[91]

In particular, the civil engineer Göran Ahlström stresses that "engineers and technicians are seen as particularly important for industrial growth because they play a crucial role in applying science and inventions in productions".[92] He underlines that

> [a] thorough theoretical and practical technical education is necessary and has been so at least from those nineteenth-century years when the science-based industries generally assumed a leading position in the industrial sector of the economies.[93]

---

[89] R. U. Mendoza, H. J. McArthur and A. O. Lopez, "Devil's Excrement or Manna from Heaven? A Survey of Strategies in Natural Resource Wealth Management", Asian Institute of Management Working Paper, 25 April 2012, p. 8.

[90] G. B. Havro, and J. Santiso, "Rescuing the Resource Rich: How Can International Development Policy Help Tame the Resource Curse?" *Working Paper Series*, June 2010.

[91] P. Lundgreen, "Engineering Education in Europe and the U.S.A., 1750–1930: The Rise to Dominance of School Culture and the Engineering Professions", *Annals of Science*, vol. 47, Issue 1 (1990), pp. 33–75.

[92] G. Ahlström, "Technical Competence and Industrial Growth", *Lund Papers in Economic History*, No. 14, (1992), p. 1.

[93] G. Ahlström, *Engineers and Industrial Growth* (London & Canberra, 1982), p. 94.

Specifically, technical education is considered one of the primary causes for Germany taking the leading role in science-based and large-scale industries from the late nineteenth century. A number of comparative studies suggest that this country developed significantly from the late nineteenth century and caught up with Britain during what is called the "second industrial revolution", likely due to a highly developed technical education system.[94] In the same line of argument, England's relative economic lag from the late nineteenth century is explained by the country's less developed education system and the scarcity of technically skilled technicians and engineers.[95]

Furthermore, specific types of education are shown to play a key role in the development of natural resource industries. David and Wright find that the development of North American mining depended on detailed scientific research, knowledge in geology, mineralogy and chemistry, and engineering work. The use of highly qualified scientists and engineers was important: "Scientifically trained personnel were instrumental in maintaining the flow of new mineral discoveries in America, through the application of increasingly sophisticated forms of geological knowledge and search procedures." This was possible, according to them, because the United States had more than 20 schools which granted degrees in mining and in 1893 the country "had more mining students than any country in Europe, except Germany".[96]

There are strong indications that basic and technical education has been vital for industrial development, but analyses tend to show a correlation between schooling, literacy, and economic growth, but do not link education directly to economic growth. And, perhaps more importantly, analyses which explain why, or how, specific types of learning and study programmes were important for innovation are scarce. Economist Richard Easterlin shows a correlation between a high education level and

[94] G. Ahlström, *Engineers and Industrial Growth* (London & Canberra, 1982); R. Fox, A. Guagnini, *Education, technology, and industrial performance in Europe, 1850–1939* (Cambridge, 1993); G. W. Roderick and M. D. Stephens, *Scientific and Technical Education in Nineteenth-Century England* (Newton Abbot, 1972).

[95] M. Sanderson, *The Missing Stratum Technical School Education in England 1900–1990s* (London, 1994), p. 145.

[96] P. David, G. and Wright, "Increasing Returns and the Genesis of American Resource Abundance", *Oxford Journals*, vol. 6, issue 2 (1997), pp. 231–234.

strong economic growth, but he focuses less on how schooling and high literacy were used in practice.[97] Much in the same line of argument, Lars Sandberg shows that there is a correlation between countries with high literacy in 1850 and the ones with high income per capita in 1970, but there is less evidence of a direct causal link to innovation.[98] He emphasises the importance of schooling and literacy in Swedish industrial and economic growth before the First World War.[99] Other scholars consider the effect of general schooling and literacy on economic growth to be an exaggeration. Joel Mokyr, in particular, does not emphasise literacy in his explanation of the causes of the British Industrial Revolution:

> Overall literacy rates can hardly have mattered as much as they are believed to in modern economies, given Britain's backwardness in that dimension. The quality of the average worker may have mattered much less for the generation and adoption of new and more productive techniques than the quality of the skilled artisans and mechanics.[100]

On the eve of Britain's industrialisation, the adult literacy rate was between 50 and 70 per cent, which was lower than that of other European countries.[101] Scandinavia, for example, represented countries where the population became literate early, even before Britain, but that industrialised much later.[102] The economic historian Kevin O'Rourke and economist Jeffrey Williamson also downgrade schooling in their analysis of Sweden. According to them it was only "modestly important" to the catching up of Sweden and Scandinavia between 1870 and the First World War. They conclude that "while schooling certainly helped make

[97] R. A. Easterlin, "Why Isn't the Whole World Developed?", *Journal of Economic History*, Vol. 41, Issue 1 (1981), p. 7.

[98] P. Sandberg, "Ignorance, Poverty and Economic Backwardness in the Early Stages of European Industrialization: Variations on Alexander Gerschenkron's Grand Theme", *Journal of economic history*, vol. 11 (1982).

[99] P. Sandberg, "The Case of the Impoverished Sophisticate: Human Capital and Swedish Economic Growth Before World War 1", *Journal of Economic History*, vol. 49, No. 1 (1979), pp. 225–241.

[100] J. Mokyr, "Knowledge, Enlightenment, and the Industrial Revolution: Reflections on *Gifts of Athena*", *History of Science*, vol. 45, part 2, (June 2006), p. 1.

[101] C. M. Cipolla, *Literacy and Development in the West* (Baltimore, 1969), p. 114.

[102] J. Mokyr, *Gifts of Athena*, (Princeton and Oxford, 2005), p. 292.

the late nineteenth century Scandinavian catch up possible, it was not the central carrier implied by so much of the literature".[103]

Much in the same way as basic education, the link between technical education and economic growth is still subject to debate. Although it is widely assumed that technical education is an important requirement for economic growth, the studies are suggestive, and there are few, if any, clear connections between technical education and industrial performance. Some find that education contributes primarily to critical thinking and that the need for higher education in the workplace is exaggerated.[104] The economist Michael Spence even suggests that there are no apparent relations between formal education and increased productivity, and that education is just a way for the job applicant to "signal" to the employer about his or her abilities. The employer assumes that there is a correlation between the qualifications of the employee and having greater skills.[105] Other studies assume that "the more engineers and technicians, the better". The general argument is that rich and developed countries have a high number of engineers, technicians, or architects, while poor and underdeveloped countries have fewer of them. Much of the research of technical education and economic growth is focused on the supply and demand of education from a quantitative point of view and they base their analyses on a number of universities and schools providing engineering and technical studies. Sometimes the number of students or graduates is compared to the total number of workers or a population.[106] For example, G. W. Roderick and M. D. Stephens—starting with the common supposition that Germany took the lead in industrial growth from the late nineteenth century—determine that the number of scientists and technicians was around five times more in Germany in 1900 compared to that

---

[103] K. O'Rourke and J. Williamson, "Education, Globalization and Catch-Up: Scandinavia in the Swedish Mirror", *Scandinavian Economic History Review*, XLIII, 3 (1995), p. 309.

[104] P. Lundgreen "Engineering Education in Europe and the U.S.A., 1750–1930: The Rise to Dominance of School Culture and the Engineering Professions", *Annals of Science*, 47, 1990, p. 34.

[105] M. Spence, "Job Market Signaling", *The Quarterly Journal of Economics*, vol. 87, No. 3 (1973), pp. 335–374.

[106] G. Ahlstrøm, "Technical education, engineering, and industrial growth: Sweden in the nineteenth and early twentieth centuries" in R. Fox, A. Guagnini, *Education, technology, and industrial performance in Europe, 1850–1939* (Cambridge, 1993), p. 122.

in England.[107] In the case of Australia, a drastic increase in the number of engineers is used to explain how the mining sector grew in the beginning of the first part of the twentieth century. Ferranto and colleagues state that this country "lagged behind the United States until after 1920 – with 47 engineers per 100 000 people to 128 per 100 000 – but would reach 163 by 1955".[108] Chile had six engineers per 100,000 workers compared to 84 in Sweden in 1890 and 128 in the United States in 1920.[109] More widely, the ratio of engineers and architects per worker is used to explain differences in growth of different countries.[110]

The functions, uses, and importance of skilled workers, engineers, and professionals, however, are not described in such analyses. The numbers of engineers or architects do not give us any information about how and why they were important. Robert R. Locke sums up studies on education and industrial performance in the following way:

> [A] closer inquiry has shown that this imposing body of knowledge, if of especial value to the study of technical education, has not treated higher education and entrepreneurial performance adequately. Actually it has, with few exceptions, not considered the subject at all. Scholars have, for the most part, discussed higher education, even in the technical and commercial sciences, comprehensively, that is, they have regarded it as part of the history of engineering or applied science or as a particular aspect of the general social and institutional transformation of a country. Rarely, moreover, have they engaged in comparative histories, for the literature is overwhelmingly nationally oriented. As for the two variables, education and entrepreneurial performance, education historians have just assumed that a

[107] G. W. Roderick and M. D. Stephens, *Scientific and Technical Education in Nineteenth-Century England*, (Newton Abbot, 1972), p. 16.

[108] D. De Ferranti et al. *From Natural Resources to the Knowledge Economy* (Washington, DC, 2002), pp. 58–59.

[109] W. F. Maloney, "Missed Opportunities: Innovation and Resource-Based Growth in Latin America", *World Bank Policy Research Working Paper* 2935 (2002), 37.

[110] See W. F. Maloney, "Missed Opportunities: Innovation and Resource-Based Growth in Latin America", *World Bank Policy Research Working Paper* 2935 (2002); F. Valencia Caicedo and W. F. Maloney, "Engineers, Innovative Capacity and Development in the Americas", *Policy Research working paper*, no. WPS 6814. Washington, DC: World Bank Group (2014).

relationship existed between them and then proceeded to demonstrate how both French and British higher education were deficient.[111]

The different findings in the literature suggest that the specific role of education and engineers in industrial development should be analysed in more detail. According to Locke, cataloguing number of students, rates of growth, and so on accounts for only part of the story:

> To determine comparative educational effectiveness, from the entrepreneurial viewpoint, one needs (aside from data on numbers, size of student body, rates of growth of schools, and background of students) information about the conditions of entrepreneurship within an economy and about the extent to which the educational institutions were able to satisfy these conditions. It cannot be assumed [...] that schools in countries did not differ significantly in curricula, instruction, and in the way they were associated with business and industry even if each country had the same number of schools and they were of equal size. An investigation into the curricula of competing national systems of higher education [...] is crucial for an analysis of higher education and entrepreneurial performance.[112]

The argument made by Wolfgang König is relevant here:

> So long as we lack research on the careers of the technical intelligentsia or on the selection processes and recruitment of industrial engineers, the question regarding the relations between technical education and economic performance cannot be answered properly.[113]

In line with the argument that knowledge is key to developing dynamic and productive natural resource–intensive industries, and that education supports learning, it seems relevant to analyse the functions of engineers and technicians in more detail.

---

[111] R. R. Locke, *The End of the Practical Man* (London, 1984), p. 2.

[112] R. R. Locke, *The End of the Practical Man* (London, 1984), p. 5.

[113] W. Kőnig, "Technical education and industrial performance in Germany: a triumph of heterogeneity", *Education, Technology and Industrial Performance in Europe, 1850–1939,* in R. Fox and A. Guagnini (eds) (Cambridge, 1993), p. 81.

Sources and Methodology

This book is an edited and revised version of my PhD thesis called "The mining sectors in Chile and Norway from ca. 1870 to 1940: the development of a knowledge gap", defended in September 2015 at the University of Geneva. The analysis focuses here mainly on the period between 1860 and 1940, which was the period before—and during—which Norway's economy started to grow faster. From the 1930s to the 1940s and onwards, the economy grew faster than it had before. The country's high economic performance is therefore not a new phenomenon and cannot solely be explained by the recent development of the oil and gas industry, which began in 1969.

This book analyses one natural resource sector, namely mining—instead of the whole economy—which allows for a more detailed empirical analysis than on the country level. Focusing on only one part of the economy, and conducting analyses of *one* industry alone, enables a more in-depth study of how innovation has occurred, and could provide important insights into how natural resource industries in some countries have become highly innovative, dynamic, and productive, while others have been less innovative. The period between 1860 and 1940 is chosen because mineral and metal production expanded considerably in Norway during these years. Moreover, world mining went through radical technological changes in this period. To continue rational operation, new energy sources were brought into use, large-scale production was adopted, and both the organisation of mining businesses and the techniques for finding, removing, and processing ores became increasingly specialised. This allows us to ask how Norway coped in a situation of technological turbulence.

The mining sector in Norway was not nearly as big as in other countries. It was instead one of many export sectors, and it accounted for around 30 per cent of exports in the early twentieth century.[114] We cannot fully establish the determinants of success in Norway by looking to one sector. The country developed multiple relatively big, and growing,

[114] SSB, *Nasjonalregnskap 1900–1929*; Statistisk sentralbyrå, *Nasjonalregnskap 1930–1939 and 1946–1951.*

natural resource industries, as well as some linked industries, in addition to mining during the nineteenth and twentieth centuries, notably fishing, agriculture, timber, and timber-related—and later oil and gas—industries. Shipping also represented an important economic sector. This made the Norwegian economy diverse and less dependent on one industry and less vulnerable. The boost in the economy from the 1930s onwards seemed to be caused by growth in several industries, and the growth in the mining sector—in particular the large-scale electro-metallurgical industry which grew from the early twentieth century—is one amongst multiple factors which explains the strong economic growth that occurred in Norway at the time. The book hypothesises that mining was one of many dynamic, innovative natural resource sectors and that this analysis exemplifies and contributes to the explanation of the development path Norway has taken to become the knowledge-intensive and high-performing economy that it is today.

Research Approach: An In-Depth Investigation of Knowledge Organisations

Is there a way to identify and analyse how knowledge has developed and how learning and innovation processes have taken place? Charles Edquist finds that "the best way of doing this is by actually using the approach in empirical (and comparative) research".[115] In the same line of argument, Bruland and Smith state that

> [i]f economic growth essentially reflects changing technological capability, and capability reflects learning and knowledge accumulation, then the key questions about Nordic growth concern how knowledge was acquired, diffused, and used.[116]

---

[115] C. Edquist, "The Systems of Innovation Approach and Innovation Policy: An account of the state of the art" (Aalborg, 2001), p. 3.

[116] K. Bruland and K. Smith, "Knowledge Flows and Catching-Up Industrialization in the Nordic Countries: The Roles of Patent Systems", in H. Odagiri et al. *Intellectual property rights, development and catch-up an International comparative study* (Oxford, 2010), pp. 63–94.

Economic historians have recently addressed such questions from a historical perspective. Instead of analysing preconditions and incentives for innovation and making economic growth models which generalise reality, scholars such as Joel Mokyr, Kristine Bruland, and Patrick O'Brien identify and analyse the actual institutions and organisations through which knowledge was created, transferred, used, and modified, and seek to explore how learning and innovation processes took place.[117] Fundamental to their approach is the notion that knowledge creation and the building of knowledge societies were based on certain "knowledge institutions" and "knowledge organisations", including scientific and technical societies, education systems, industrial exhibitions, industrial espionage operations, study trips for learning, research centres, and business firms, all of which interacted and collaborated with each other and actively and directly enabled innovation. In his analysis, Joel Mokyr emphasises that such institutions and organisations, and their collaborations and interactions, were in the core of what he calls the "Industrial Enlightenment". He writes that

> [t]he economies that were most successful in the second Industrial Revolution were those in which the connections were most efficient. The institutions that created these bridges are well understood: universities, polytechnical schools, publicly funded research institutes, museums, agricultural research stations, research departments in large financial institutions.[118]

Similar types of knowledge institutions and organisations are explored for the Nordic countries and used to explain their radical economic

[117] See, for instance, J. Mokyr, *Gifts of Athena*, (Princeton and Oxford, 2005); J. Mokyr, *The Enlightened Economy* (New Haven, 2009); K. Bruland, "Kunnskapsinstitusjoner og skandinavisk industrialisering" in *Demokratisk konservatisme,* Engelstad, F. and Sejersted, F. (eds.) (Oslo, 2006); K. Bruland, "Skills, learning and the International Diffusion of Technology: a Perspective on Scandinavian Industrialization", in *Technological Revolutions in Europe,* M. Berg and K. Bruland (eds.) (Cheltenham, 1998); K. Bruland and K. Smith, "Knowledge Flows and Catching-Up Industrialization in the Nordic Countries: The Roles of Patent Systems", in *Intellectual property rights, development and catch-up an International comparative study,* H. Odagiri et al. (Oxford, 2010), pp. 63–94; P. O'Brien, *Stages in the Evolution of a Western regime for the discovery, development and diffusion of useful and reliable knowledge*, URKEW *Discussion Paper* No. 7, London School of Economics (2011).

[118] J. Mokyr, *Gifts of Athena* (Princeton and Oxford, 2005), p. 102.

growth during the past two centuries. Economists and historians find multiple channels for knowledge transfer, diffusion, and adoption. Technical and scientific societies, for example, often published technical magazines and journals, and they are found to be important channels for knowledge diffusion in the eighteenth and nineteenth centuries. In such publications, engineers and professionals obtained information about new and existing industrial techniques and methods and descriptions of their use. Their function was, in particular, to spread information and publish updates on new technologies and patents. Many industrialists and engineers were members of technical and scientific societies and had access to such journals. It is also found that technical and scientific societies, in addition to making publications, functioned as meeting places for exchange of relevant and useful knowledge.[119]

Education systems may have been particularly important for innovation and industrial performance in the Nordic countries, but we lack detailed empirical analyses. One of the main purposes of schools, universities, and technical schools in general has been to facilitate learning. In the Nordic countries, school systems developed early, and their populations were some of the first in Europe to become literate, which in turn is considered important for the spread of general knowledge and awareness in countries.[120] Professor Peter William Musgrave has perhaps made the most in-depth detailed comparative analysis of education and industrial development. This study differs from other analyses in that the use of technology and knowledge requirements are accounted for before considering the actual supply of professionals. Musgrave detects the use of specific skills which appeared to be important for the British and German iron and steel industries from the mid-nineteenth century to the mid-twentieth century and relates them to particular knowledge domains and qualifications. As an example, he finds that

[119] K. Bruland and K. Smith, "Knowledge Flows and Catching-Up Industrialization in the Nordic Countries The Roles of Patent Systems", in H. Odagiri et al. *Intellectual property rights, development and catch-up an International comparative study* (Oxford, 2010), pp. 63–94; K. Bruland, "Skills, learning and the International Diffusion of Technology: a Perspective on Scandinavian Industrialization", in M. Berg and K. Bruland (eds.), *Technological Revolutions in Europe* (Cheltenham, 1998), p. 177.

[120] P. Sandberg, "The Case of the Impoverished Sophisticate: Human Capital and Swedish Economic Growth Before World War 1", *Journal of Economic History*, vol. 49, No. 1 (1979), pp. 225–241.

changes in technology were based on new capabilities and specialisations from the 1860s:

> The growing stress on science was leading to the employment of chemists in the industry. Here can be seen the beginning of the grade of "technician". Another growing point was associated with the expansion of the counting house, calling for more clerks whose education ranked them in this grade. The export trade would demand some knowledge of foreign language. Such men needed a broad education with specific attention to science and mathematics for potential works chemists.[121]

At the same time, it is important to consider that there have been other ways of obtaining knowledge than from formal theoretical schooling and scientific research. This is emphasised by Joel Mokyr, yet empirical studies on this subject are also scarce. He points out that skills and talents in eighteenth-century Britain were created as much by practical experience and commercial culture as through formal education.[122] A vital point made in the literature of the Nordic countries is that innovation has largely been based on knowledge transfers from other countries. Visiting industrial exhibitions were one way through which workers, technicians, and engineers acquired contacts and practical knowledge about new techniques. Industrial exhibitions were held in a number of European cities during the nineteenth century and were crucial in presenting "the state of the art" in mechanical and industrial methods.[123] Bruland, for example, finds that visits to exhibitions led to the import of reaping equipment, which was diffused in Norway via local farm fairs.[124] Göran Ahlström shows that industrial exhibitions were of great importance to the diffu-

[121] P. Musgrave, *Technical Change, the Labour Force and Education* (Oxford, New York, 1967), p. 27.

[122] See J. Mokyr, "Knowledge, Enlightenment, and the Industrial Revolution: Reflections on *Gifts of Athena*", *History of Science*, vol. 45, part 2, (June 2006).

[123] K. Bruland, "Skills, learning and the International Diffusion of Technology: a Perspective on Scandinavian Industrialization", in M. Berg and K. Bruland (eds.), *Technological Revolutions in Europe* (Cheltenham, 1998), p. 182.

[124] K. Bruland and K. Smith, "Knowledge Flows and Catching-Up Industrialization in the Nordic Countries: The Roles of Patent Systems", in H. Odagiri et al. *Intellectual property rights, development and catch-up an International comparative study* (Oxford, 2010), pp. 63–94.

sion of technology and knowledge more generally in the Nordic countries.[125] The underlying idea here is that participating in exhibitions would contribute to transfer of technology by facilitating an awareness of new equipment, machinery, and working methods which would increase efficiency in industries.

More widely, "study trips" abroad were found to be highly important for practical learning, and in turn for capacity building and industrial development. The workshop and textile industries in nineteenth-century Norway, for example, are found to be built largely on practical technical experience from companies and organisations abroad.[126] Longer work experience from other countries also seemed to be important. Per-Olof Grönberg finds work experience from European and North American countries to be highly valued amongst Swedish firms. In his analysis of Swedish engineers from 1880 to 1930, he shows that they were recruited after their return and describes how they used their experience from abroad directly in work situations.[127]

Such approaches, which aim at exploring transfer and use of knowledge, and channels through which learning occurred and knowledge was accumulated and used seem highly useful to further our understanding of underlying factors of industrial development and economic growth, although much remains to be done. Edward Glaeser points out that the hypothesis of links between the Industrial Revolution and the thinkers of the Enlightenment is not really tested.[128] Studies which empirically and systematically analyse where knowledge came from, how it was diffused, and the people involved in these processes are scarce. More widely, it is emphasised by many that, although there is an increased focus on such questions now, we still know little about how knowledge has been accumulated

[125] G. Ahlström, *Technological Development and Industrial Exhibitions 1850–1914* (Lund, 1995), p. 11.

[126] K. Bruland, "Norsk mekanisk verkstedindustri og teknologioverføring 1840–1900", in *Teknologi i virksomhet: verkstedindustri i Norge etter 1840*, E. Lange (ed.) (Oslo, 1989), pp. 46–47; K. Bruland, *British technology and European industrialization* (Cambridge, 1989).

[127] P-O. Grönberg, *Learning and Returning. Return Migration of Swedish Engineers from the United States, 1880–1940* (Sweden, 2003), p. 1.

[128] Edward Glaeser, Thinkers and Tinkerers, History, New Republic, 22 June 2010.

and how organisational and institutional structures have influenced innovation directly in Britain, Europe, but also other parts of the world.[129]

A key point to be made here is that to understand differences in development across natural resource–intensive economies, and the foundation of economic growth, historical analyses of learning and innovation which enabled the utilisation of natural resources are essential. Drawing on these approaches, the idea here is to go beyond general categorisations of countries and take into account the complexity of development processes, but at the same time to explore systems by which learning happens and knowledge is used. I investigate mining development in Norway by analysing empirically how certain knowledge organisations at different levels created, transferred, diffused, modified, and used knowledge and how knowledge was transformed by learning into technological innovation. This is done by looking to the past. Taking into consideration the path dependence of institutions, the question here is whether such high-performing natural resource industries—mining in particular—have a longer historical tradition in Norway, and, if so, whether knowledge-enhancing organisations have played a role. As a starting point, we know that multiple public and private mining knowledge organisations which aimed to develop scientific and technical knowledge for mining developed in Norway from early on (Table 1.1).

Yet, the mere detection of these initiatives does not tell us anything about their activities, knowledge development, outcomes, and links to innovation. We still do not know whether they influenced learning and innovation processes. Therefore, to learn more about them, I aim to analyse them in more detail. Based on the argument that learning depends on an innovation-friendly institutional and organisational context, I explore their activities and performances.

---

[129] See N. Craft, "Explaining the first Industrial Revolution: two views", *European Review of Economic History*, vol. 15 (1), 2011; C. Edquist, "The Systems of Innovation Approach and Innovation Policy: An account of the state of the art" (Aalborg, 2001); K. Bruland, *British technology and European industrialization* (Cambridge, 1989).

Table 1.1 Selected knowledge organisations in Norway aimed to develop knowledge for mining

Organisation	Aims	Year of establishment
Domestic mining companies	Private or public business organisations which aimed to produce metals and minerals and sell them for profit	Continuously
Foreign mining companies	Foreign business organisations, normally on a large scale, aimed to produce metals and minerals for profit	Continuously
Mining education on tertiary level	Organisations which aimed to provide high-quality mining instruction, with emphasis on scientific and theoretical courses	1757
Mining education on intermediate level	Organisations which aimed to provide high-quality mining instruction, with more emphasis on practice	1866
"The Norwegian Mining Engineer Association"	Interest group which aimed to promote mining	1900
Journal of Mining	Mining journal which published information and news about mining in Norway and abroad	1913
Scholarships for study travels and working experience abroad	Private and public funds for technicians and engineers to travel abroad and acquire practical experience, notably of foreign technology	From late eighteenth century
Industrial exhibitions (participation)	Organised presentation of new mining technology	From mid-nineteenth century
"Geological Survey of Norway"	Research centre which aimed at making geological maps, finding prospects, and analysing ore for metal and mineral mining and extraction	1858
Mechanical workshops	Private firms which aimed to supply mining companies with inputs of mechanical devices, machinery, technical services, tools, etc.	From the 1840s

Concepts and Definitions

Mining

"Mining" is defined widely as the production of metal ores, non-metallic minerals, stones, rocks, and energy minerals.[130] More specifically, mining involves the removal of metal and mineral ore from the earth along with the processing of metals and minerals. In order to obtain the desired product, normally in as pure a state as possible, a number of procedures are required. First, the ore deposit needs to be found and geological mapping and surveys carried out to investigate conditions beneath a given ground. Subsequently, the ore is dug out and removed from the earth. After the ore is removed and transported from the mine the mineral is processed so that the useful mineral can be separated from the worthless rock mass. Big pieces of material are first crushed into smaller chunks or powder through crushing and milling. As much gangue as possible is removed before smelting. Minerals are then separated from each other through a variety of methods depending of the physical and chemical characteristics of the composition of the ore. Finally, pure metals and minerals are produced using a range of smelting and refining methods. "The mining sector in Norway" includes here all metals and minerals that were produced in Norway and are included in the official statistics.[131]

Technology, Technology Transfer and Innovation

Here, the broad and subdivided definition of technology presented by Bruland and Mowery is used. They divide technology into "knowledge", "techniques", and "organisation" to capture the factors that have enabled production. First, knowledge entails the understanding and skills which are necessary for production. This involves scientific knowledge and natural laws as well as engineering, know-how, and operative skills. In mining, this would refer to the underlying knowledge used to measure,

[130] H. Carstens, ...*Bygger i Berge: en beretning om norsk bergverksdrift* (Trondheim, 2000), 10.

[131] Statistisk sentralbyrå, *Norges offisielle statistikk, Norges Bergverksdrift* (Oslo, 1866–1940).

design, and build the tunnels, adits, shafts, and so on of mines, and to remove, transport, and process ore. Second, technology further involves techniques, that is the machines, tools, and equipment, and the instructions and processes required to employ, repair, and maintain these. In mining, these were, for example, steam engines and turbines used for lifting and drainage, mechanical and electric equipment used inside and outside the mines, and crushers, converters, and furnaces used in the processing of metal and mineral ores. Third, organisation involves the administration, management, and coordination systems through which productions occur. In mining, they include the management of workers and equipment, as well as the administration of the many mining divisions, which include mines, crushing and smelting plants, laboratories, and workshops, amongst others. These three aspects of technology are integrated in production processes.[132]

The economist Richard Nelson defines innovation as "the process by which new products and techniques are introduced into the economic system". Innovation is often divided into product and process innovation. A process innovation involves the introduction of a new hardware or changes in the way equipment, machinery, and devices are used. Such changes involve modifications in operating processes. Product innovation involves the introduction of a completely new product. A successful innovation, defined by Nelson, is a change which results in the capability of doing something that is not previously done, or at least not so well, or so economically.[133] Analyses of innovation are challenging because there is no general prototype for how technological changes actually occur. One of the problems is that innovations happen randomly and without any specific order. They are regarded as processes rather than isolated events. Introducing new and more efficient working methods, or improving a product, is subject to a number of different experiments and testing, and vary from case to case. Kline and Rosenberg state that "the processes and systems used are complex and variable; that there is no single correct

[132] K. Bruland and D. C. Mowery, "Technology and the spread of capitalism", in *Cambridge History of Capitalism, vol. II,* L. O'Neil and J. G. Williamson (Cambridge, 2014), 84.

[133] R. R. Nelson, "Innovation", in *International Encyclopedia of social sciences* 7 (1968), pp. 339–345.

formula, but rather a complex of different ideas and solutions that are needed for effective innovation".[134] They stress that

> [i]nnovation is complex, uncertain, somewhat disorderly, and subject to changes of many sorts. Innovation is also difficult to measure and demands close coordination of adequate technical knowledge and excellent market judgment in order to satisfy economic, technological, and other types of constraints – all simultaneously. The process of innovation must be viewed as a series of changes in a complete system not only of hardware, but also market environment, production facilities and knowledge, and the social contexts of the innovation organization.[135]

One of the key challenges when studying innovation is to identify how learning is transformed into technological change. Simple modifications, such as new ways of using a device, are often harder to detect than the adoption of a new machine. They are, nevertheless, important. Bruland and Mowery confirm that "radical" innovations

> typically assume their economic importance as a result of numerous, individually modest but cumulatively significant, "incremental" improvements after the introduction of a new product. The importance of incremental innovation also underscores the complex relationship between the appearance of a new technology and its adoption.[136]

Bruland shows how complex innovation processes can be in her study of the building of the textile industry in Norway in the mid-nineteenth century. The development of this industry involved multifaceted technology transfers from Britain in collaboration with local workers. Not only were equipment and machinery imported to enable successful operations, but also knowledge, expertise, and information were transferred

---

[134] D. Kline and N. Rosenberg, "An Overview of Innovation", in R. Landau and N. Rosenberg (eds.) *The Positive Sum Strategy: Harnessing Technology for Economic Growth* (Washington, DC, 1986), p. 279.

[135] D. Kline and N. Rosenberg, "An Overview of Innovation", in R. Landau and N. Rosenberg (eds.) *The Positive Sum Strategy: Harnessing Technology for Economic Growth* (Washington, DC, 1986), p. 275.

[136] K. Bruland and D. C. Mowery, "Technology and the spread of capitalism", in *Cambridge History of Capitalism, vol. II*, L. O'Neil and J. G. Williamson (Cambridge, 2014), p. 84.

through travels abroad, and used in operations with help from foreign skilled workers and labour.[137] It is therefore important to consider all kinds of changes in technology, both large and small.

An important factor which should be stressed here is that Norway was a "catching-up economy" of the Industrial Revolution and new technology often had its origin in industrial powers, such as Britain, Germany, France, and the United States. Innovation, therefore, largely involved "technology transfer" from these countries, which meant transmitting techniques, methods, and systems from other countries and adapting them to local conditions and using them efficiently in operation. Small follower countries were usually not able to make investments, research, and development of similar magnitude as large economies, which made their capacity to absorb, adapt, and use foreign knowledge essential to their economic performances. Bruland points out that in the case of small countries—such as Norway—"[w]hether they undergo industrialisation processes and remain in the forefront in terms of advanced industrial performance depends largely on whether they can develop the ability to apply technologies developed abroad".[138] In mining, technology transfer would involve the allocation of mining techniques, ore dressing and smelting methods, business systems, and so on from one place to another and the imitation, modification, adaptation, and preparation it involved in using them successfully in operation.

Knowledge Organisations and Institutions

Technology transfer and innovation processes do not occur in isolation. They occur in an institutional setting. The Nobel Laureate in economics Douglass C. North makes a distinction between institutions and organisations. "Institutions" are understood as "rule of the game" and are frameworks of formal written laws, norms, codes of conduct, government policies, and so on which regulate the economic behaviour of individuals,

[137] K. Bruland, *British technology and European industrialization* (Cambridge, 1989).

[138] K. Bruland, "Norsk mekanisk verkstedindustri og teknologioverføring 1840–1900", in *Teknologi i virksomhet: verkstedindustri i Norge etter 1840*, E. Lange (ed.) (Oslo, 1989), p. 34.

firms, and other organisations. "Organisations" are groups of individuals bound by a common purpose to achieve objectives. They are players or actors.[139] These wide definitions of institutions and organisations are useful, since, in principle, they cover all activities related to innovation processes. The framework of rules and norms in which actors operate is the core of economic activity, and it is how entrepreneurs, company managers, workers, engineers, politicians, and others operate which determines the economic result.

"Knowledge organisations and institutions" are here understood as institutions and organisations which regulate, obtain, accumulate, adopt, use, modify, transfer, and diffuse knowledge. Knowledge institutions include, for instance, economic regulations and arrangements for the spread of knowledge, government policies for travel and immigration, as well as educational systems. Knowledge organisations include, for instance, scientific and technical societies, educational establishments, research centres, industrial exhibitions, as well as business firms.[140]

Learning and Knowledge

Technological change involves learning. Actors have to acquire the skill, or capability, to do something new in order to innovate. The process of learning in turn means acquisition of knowledge.[141] Knowledge is here subdivided into different categories. Knowledge used for productive purposes often includes both "codified" and "tacit" knowledge. Codified knowledge is knowledge which has been articulated, written down, and stored. On the other hand, there is a certain knowledge which is more difficult to pass on to others. In the case of mining, the uniqueness of

[139] D. C. North, "Institutions" in *The Journal of Economic Perspectives*, Vol. 5, No. 1, (1991), p. 97.

[140] J. Mokyr, *Gifts of Athena*, (Princeton and Oxford, 2005); J. Mokyr, *The Enlightened Economy* (New Haven, 2009), K. Bruland, "Kunnskapsinstitusjoner og skandinavisk industrialisering" in *Demokratisk konservatisme,* Engelstad, F. and Sejersted, F. (eds.) (Oslo, 2006); K. Bruland, "Skills, learning and the International Diffusion of Technology: a Perspective on Scandinavian Industrialization", in M. Berg and K. Bruland (eds.), *Technological Revolutions in Europe* (Cheltenham, 1998).

[141] K. J. Arrow, "The economic implications of learning by doing", *The Review of Economic Studies,* Vol. 29, No. 3 (1962), p. 155.

each working plant means that this type of industrial activity has involved a large degree of trial and failure. The nature of mining indicates that mining activities have largely involved physical and practical tasks of planning, drawing, designing and making mines, removing ore by digging, carrying, and transporting, and finally making pure metal and mineral products through washing, crushing, separating, and smelting. The ore needs to be tested and analysed, mine constructions have to be evaluated and remade, and working techniques have to be tested, selected, and modified. The physical and practical characteristics of this industry suggest that the activities, to a large degree, have been based on "tacit knowledge", which is defined by Joel Mokyr as "implicit skills such as dexterity, hand-eye coordination, and sense of 'what worked'".[142] According to the scientist Michael Polanyi tacit knowledge is knowledge which can hardly be described or explained. He states that "we can know more than we can tell".[143] Nelson and Winter further explain that "the knowledge that underlies a skilful performance is in a large measure tacit knowledge".[144] In particular, the tacit dimension has been pointed out by Joel Mokyr to be particularly important in the case of mining. He stresses that the "centrality of tacit knowledge [...] was especially true for the metal and coal industries".[145]

Since tacit knowledge cannot be expressed, it often passes on from person to person. It is acquired through observation or practice in a specific context through "learning by observing" or "learning by doing". Nobel Laureate in economics Gary S. Becker explains that the adequate way of learning depends on the nature of the knowledge:

> Some type of knowledge can be mastered better if simultaneously related to a practical problem; others require prolonged specialization. That is, there are complementary elements between learning and work and between learning and time. Most training in the construction industry is apparently

[142] J. Mokyr, *Gifts of Athena*, (Princeton and Oxford, 2005), p. 73.

[143] M. Polanyi, *The Tacit Dimension* (Gloucester, 1983), p. 4.

[144] R. R. Nelson and S. G. Winter, *An Evolutionary Theory of Economic Change* (Cambridge, 1982), p. 73.

[145] J. Mokyr, "Knowledge, Enlightenment, and the Industrial Revolution: Reflections on *Gifts of Athena*", *History of Science*, vol. 45, part 2, (June 2006), pp. 1–2.

> still best given on the job, while the training of physicists requires a long period of specialized effort. The development of certain skills requires both specialization and experience and can be had partly from firms and partly from schools...[146]

For instance, words are viewed as a more effective tool when learning elementary algebra than when learning carpentry.[147] To understand how to use, maintain, and repair a tool, a drill, an engine, or a machine, reading about them in a book is not sufficient. They should be used and practised with, probably multiple times, before they can be operated with success. Polanyi explains that when a tool is used "[w]e are attending to the meaning of its aspect on our hands in terms of its effect on the things to which we are applying it".[148] It is therefore essential to consider other arenas for learning than schools and classrooms. The capacity to solve practical problems is often acquired from experience, usually in working situations. Kenneth Arrow, for example, finds that workers solve problems and daily tasks, and become more efficient, through learning by observing and learning by doing.[149] Following this line of argument, there are some skills that cannot be learned in a school classroom, but which are instead acquired in a practical work setting.

Sources and Chronology

Chapter 2 gives an outline of productions, exports of minerals and metals, productivity, and technological development from approximately 1860 to 1940. The empirical analysis is divided into two main parts. Chapter 3 provides a framework for the types of knowledge and learning processes on which mining in Norway was based. This framework functions as a starting point for the subsequent analysis of knowledge organisations. The

[146] G. S. Becker, *Human Capital* (New York, 1975), p. 37.

[147] R. R. Nelson and S. G. Winter, *An Evolutionary Theory of Economic Change* (Cambridge, 1982), p. 78.

[148] M. Polanyi, M. *The Tacit Dimension* (Gloucester, 1983), p. 13.

[149] K. J. Arrow, "The economic implications of learning by doing", *The Review of Economic Studies*, Vol. 29, No. 3 (1962).

framework does not claim to present rigid and linear systems of knowledge bases, but rather includes key knowledge aspects involved repeatedly in the development of innovative, dynamic, and technologically experimental mining projects. Chapters 4, 5, 6, and 7 investigate four key knowledge organisations—educational establishments, mining companies, mechanical workshops, and the Geological Survey of Norway (NGS) of Norway—and their activities, their knowledge contributions, functions, and performances in innovation processes. The variables I look at and sources I use for each organisation are the following:

1. The University, the Norwegian Institute of Technology (NIT), and the Mining School
   (a) Scientific and theoretical knowledge: study programmes and course descriptions exist for both intermediate and higher level and are used to analyse the content of the formal mining education in relation to technological changes at the time.
   (b) Mining engineers and technicians: student yearbooks include detailed information about all mining engineers graduating in Norway and abroad from 1787 to 1940 and mining technicians graduating from 1869 and 1940. With this information we can calculate how many they were and their share of total number of workers and companies.
   (c) Other trained workers: a near-complete picture of the educational background of all formally trained workers in the mining sector in Norway from 1866 to 1940, that is secondary, technical, economic, or scientific education is obtained through student yearbooks. The yearbooks are complete collections of biographies of high school and engineering graduates for the given years and include detailed information about work and work positions. With these data, we can analyse the composition of trained workers and the share of trained workers over time.
2. Mining companies: domestic and foreign businesses
   (a) Hands-on practice and work experience: student yearbooks provide unique information about the work and positions of all the mining engineers and technicians, which allows for a detailed analysis of their work practice.

(b) Knowledge networks: student yearbooks were normally made 25 years and/or 50 years after graduation, which allows detailed analyses of the mining engineer graduates throughout their whole working career. They are used here to explore networks between companies, universities, and research centres through job switching.
(c) Travel scholarships: student yearbooks also provide information about scholarship arrangements and travels and work abroad, which allows for an analysis of learning processes of the mining engineers in other countries and knowledge transfers to Norway.

3. Mechanical workshops

(a) Equipment, machinery, and technical services to mining companies: secondary literature and company records, notably of Trondhjem Mechanical Workshop and the Bede Metal and Chemical Co. Ltd., are used to explore the use of mining machinery and equipment by mining companies and the linkages between the capital goods industry and the mining sector.

4. Geological Survey of Norway

(a) Geological mapping and ore surveys: mining journals, engineering reports, and secondary literature are used to analyse the work of geological mapping and ore surveys by the NGS in Norway.

Part 4 provides a summary of the previous chapters, and also stresses the interactions and collaboration processes between the knowledge organisations and the combination of knowledge that were used. It also seeks to explain the underlying institutional mechanisms and drivers of these organisations, and the historical openness and "outward-looking" attitude in Norway.

Research Questions

The following are the main questions I ask: On what knowledge was mining in Norway based? Were knowledge organisations involved in developing knowledge for mining, and if so, how did they function and perform? The aims are to further our understanding of (1) the functions and performances of knowledge organisations, and (2) how they may contribute to explain technological changes in the mining sector. The overall purpose here is to further our understanding of how and why Norway developed into becoming one of the richest countries in the world from an economic historical perspective.

Summary

The idea that natural resource–intensive economies only experience slow growth lacks empirical evidence. Many countries rich in natural resources have stagnated and experienced slow economic growth, but some of the richest countries in the world, notably the Nordic countries, Australia, and New Zealand, have based their economies largely on natural resource industries. Similar to Latin American and African countries, their economies rest on utilisation of natural resources, such as metals and minerals, forests, fish, and agricultural production. Differences in growth between natural resource–intensive economies suggest that there is a fundamental problem with the resource curse hypothesis, which in turn seems to be related to generalisations of countries and oversimplified models.

Why have some countries rich in natural resources become rich while others have not? This question leads to another problem, which is the assumption that natural resource industries are less innovative and dynamic than manufacturing and high-tech industries. New research suggests that this is not the case. Some natural resource industries are found to be based on highly complex knowledge, intensive learning, and dynamic linkages to other industries, much like other industries. It may be, then, that the challenge for natural resource–intensive economies is not to move away from natural resource industries and develop other types of industries, but instead to encourage the utilisation of natural resources and encourage specialisations within natural resource industries. Differences between successful and unsuccessful natural resource–intensive economies seem to be linked to whether the natural resource potential in the country is benefitted extensively and used to develop productive natural resource industries, or not. The analysis by David and

Wright about North American mining, for example, shows that from the late nineteenth century until the mid-twentieth century, the United States took more advantage of its metal and mineral potential than any other country at the time. If some natural resource industries have been innovative, while others have not, which mechanisms determine growth in the successful ones?

There is a general consensus that knowledge is the underlying foundation of economic growth and that an institutional setting which encourages and facilitates innovation is key for technological knowledge accumulation and innovation. Both historical analyses and economic studies of recent decades agree that a favourable and innovative-friendly institutional setting has been indispensable for economic growth. It is how entrepreneurs, company managers, workers, traders, engineers, and so on operate and interact with each other which determines the economic result. Looking to natural resource–intensive economies in particular, there is evidence that the successful ones have actually established organisations and institutions which have encouraged and stimulated the utilisation of natural resources and the development of natural resource industries, instead of insisting to move away from them.

There is a general consensus that learning and innovation are dependent on the society and the institutional environment in which they exist; however, there are major inconsistencies in the literature about what a "favourable institutional setting" looks like. Moreover, scholars largely disagree about which institutions influence innovation and growth. Much of the literature focuses on making general models of institutions and economic growth, which are often based on only a few countries and present broad comparisons and classifications of poor and rich countries. These models ignore specificities of each country and do not reveal the complex development paths which characterise each society. Another problem is that there is scarce empirical evidence of direct links between organisations, institutions, innovation, and economic growth. This issue is related to the fact that there is a lack of understanding of *how* institutions actually influence innovation. The focus in the literature is largely on incentives and constraints of economic transactions while few attempts have been made to actually explore how learning and innovation occur and the actors involved. Certain institutions are often assumed to encourage and motivate innovation, while other institutions do not. Certain "right" institutions should be in place: defined property rights, good and extensive infrastructure, high-quality education and research institutions, free trade policy, democratic institutions, and so on. Other institutions, such as rent-seeking and non-democratic institutions, are found to be negative for economic growth. Institutional approaches which focus on incentives are extremely useful, but there are still several uncertainties and continuing problems. Incentives or motivations to invest and innovate, understood as conditions, do not themselves explain how actors have acquired and used technological capabilities. Secure property rights have probably provided the motivation to invest, efficient and clear systems for imports and exports may have facilitated trade, and institutions focusing on transparency may have motivated honest transactions, but they do not show the knowledge accumulation and the different learning processes on which innovation has been based.

Schools, universities, and technical schools aim to encourage learning and the spread of knowledge; however, although a number of studies find that education is important for the fostering of industries, few studies really link the scientific knowledge and skills to actual building of industries. Even innovation system approaches are criticised for not including formal education, and for not capturing the complexity of learning and innovation processes. To understand more about the foundation of modern economic growth, the institutions and organisation involved in the accumulation and use of knowledge to build companies and industries should be explored further.

It has now been more common amongst economic historians to seek to explore how knowledge was transferred, used, modified, and diffused and how learning was transformed into technological innovation from a historical perspective. Certain knowledge organisations and institutions, such as education and technical education systems, industrial exhibitions, technical magazines, research centres, travel funds for study travels, immigration systems to attract foreign workers, business firms, industrial societies, and research centres, are found to be directly involved in learning and innovation processes in Europe. In some countries, the state encouraged and created some of these institutions, which gave it a special role. I seek to supplement this literature by looking to the past and exploring the functions and direct and indirect links to innovation of some of these types of knowledge organisations which aimed to develop knowledge for mining in Norway.

Bibliography

Secondary Sources (Literature)

Acemoglu, D., & Robinson, J. A. (2012). *Why Nations Fail.* London: Profile Books.

Ahlström, G. (1982). *Engineers and Industrial Growth.* London/Canberra: CROOM HELM.

Ahlström, G. (1992). *Technical Competence and Industrial Growth.* Lund Papers in Economic History, No. 14.

Ahlström, G. (1995). *Technological Development and Industrial Exhibitions 1850–1914.* Lund: Lund University Press.

Andersen, A. D. (2012). Towards a New Approach to Natural Resources and Development: The Role of Learning, Innovation and Linkage Dynamics. *International Journal of Technological Learning, Innovation and Development, 5*(3), 291.

Andrade, S., & Morales, J. (2007). *The Role of the Natural Resource Curse in Preventing Development in Politically Unstable Countries: Case Studies of Angola and Bolivia.* Institute for Advanced Development Studies, Development Research Working Paper Series No. 11.

Arezki, R., & Bruckner, M. (2011). *Resource Windfalls and Emerging Market Sovereign Bond Spreads: The Role of Political Institutions.* Research Paper No. 2011-08. Australia: University of Adelaide, School of Economics.

Arrow, K. J. (1962). The Economic Implications of Learning by Doing. *The Review of Economic Studies, 29*(3).

Bakwena, M. (2010). *Do Better Institutions Alleviate the Resource Curse? Evidence from a Dynamic Panel Approach*, PhD Thesis. Queensland: The University of Queensland Publication.

Bakwena, M., et al. (2009). *Avoiding the Resource Curse: The Role of Institutions.* MRG Discussion Paper Series from School of Economics, University of Queensland, Australia No. 32. Australia: University of Queensland.

Basberg, B. L. (2006). Patenting and Early Industrialization in Norway, 1860–1914. Was There a Linkage? *Scandinavian Economic History Review, 54*(1), 4.

Becker, G. S. (1975). *Human Capital*. New York: Columbia University Press.

Berg, M., & Bruland, K. (Eds.). (1998). *Technological Revolutions in Europe*. Cheltenham: Edward Elgar.

Bergh, T., et al. (1983). *Norge fra u-land til i-land: vekst og utviklingslinjer 1830–1980*. Oslo: Gyldendal.

Bhattacharyya, S., & Hodler, R. (2009). *Natural Resources, Democracy and Corruption*. OxCarre Research Paper, No. 2009-20.

Blomström, M., & Meller, P. (Eds.). (1991). *Diverging Paths Comparing a Century of Scandinavian and Latin American Economic Development*. Washington, DC: Johns Hopkins.

Breitwieser, A., & Foster, N. (2012). *Intellectual Property Rights, Innovation and Technology Transfer: A Survey*. The Vienna Institute for International Economic Studies Working Papers, 88.

Brooks, S. M., & Kurtz, M. (2011). Conditioning the "Resource Curse": Globalization, Human Capital, and Growth in Oil-Rich Nations. *Comparative Political Studies, 44*(6), 747–770.

Bruland, K. (1989). *British Technology and European Industrialization*. Cambridge: Cambridge University Press.

Cabrales, A., & Hauk, E. (2009). The Quality of Political Institutions and the Curse of Natural Resources. *The Economic Journal, 121*, 58.

Caicedo, F. V., & Maloney, W. F. (2014). *Engineers, Innovative Capacity and Development in the Americas*. Policy Research Working Paper; No. WPS 6814. Washington, DC: World Bank Group.

Carstens, H. (2000). …Bygger i Berge: en beretning om norsk bergverksdrift. Trondheim: Norsk bergindustriforening Den norske bergingeniørforening Tapir.

CEDA. (2007, July). *Competing from Australia*, No. 58. Melbourne: Committee for Economic Development of Australia.

Cipolla, C. M. (1969). *Literacy and Development in the West*. Baltimore: Penguin Books Ltd.

Collier P., & Hoeffler, A. (2004). Greed and Grievance in Civil Wars. *Oxford Economic Papers*, 56.

Craft, N. (2011). Explaining the First Industrial Revolution: Two Views. *European Review of Economic History, 15*, 1.

Culver, W. W., & Reinhart, C. J. (1989). Capital Dreams: Chile's Response to Nineteenth-Century World Copper Competition. *Comparative Studies in Society and History, 31*(4), 722.

David, P., & Wright, G. (1997). Increasing Returns and the Genesis of American Resource Abundance. *Industrial and Corporate Change, 6*(2), 203.

De Ferranti, D., et al. (2002). *From Natural Resources to the Knowledge Economy*. Washington, DC: The World Bank.

Easterlin, R. A. (1981). Why Isn't the Whole World Developed? *Journal of Economic History, 41*(1), 1.

Edquist, C. (2001, June 12–15). *The Systems of Innovation Approach and Innovation Policy: An Account of the State of the Art*. Lead Paper Presented at the DRUID Conference, Aalborg.

Eicher, T., & Turnovsky, S. (Eds.). (2003). *Inequality and Growth: Theory and Policy Implications*. Cambridge, MA: MIT Press.

Fagerberg, J., et al. (2008, June). *Innovation-Systems, Path-Dependency and Policy: The Co-evolution of Science, Technology and Innovation Policy and Industrial Structure in a Small, Resource-Based Economy*. Working Papers on Innovation Studies 20080624, Centre for Technology, Innovation and Culture, University of Oslo.

Falvey, R., & Foster, N. (2006). *The Role of Intellectual Property Rights in Technology Transfer and Economic Growth: Theory and Evidence*. Vienna: United Nations Industrial Development Organization.

Fosu, A. K. (2013). *Development Success: Historical Accounts from More Advanced Countries*. Oxford: Oxford University Press.

Fox, R., & Guagnini, A. (1993). *Education, Technology, and Industrial Performance in Europe, 1850–1939*. Cambridge: University of Cambridge.

Frankel, J. A. (2010). The Natural Resource Curse: A Survey. NBER Working Paper 15836.

Gerschenkron, A. (1962). *Economic Backwardness*. Cambridge: Belknap Press of Harvard University Press.

Glaeser, E. (2010, June 22). Thinkers and Tinkerers. History, *New Republic*.

Greif, A. (2006). *Institutions and the Path to the Modern Economy: Lessons from Medieval Trade*. Cambridge: Cambridge University Press.

Grönberg, P.-O. (2003). *Learning and Returning. Return Migration of Swedish Engineers from the United States, 1880–1940*. PhD Thesis. Sweden: Umeå University.

Gylfason, T. (2001). Natural Resources, Education and Economic Development. *European Economic Review, 45*, 847.

Havro, G. B., & Santiso, J. (2010, June). *Rescuing the Resource Rich: How Can International Development Policy Help Tame the Resource Curse?* ESADE Knowledge, Working Paper Series.

Hirsch-Kreinsen, H., et al. (2003). Low-Tech Industries and the Knowledge Economy: State of the Art and Research Challenges. In *PILOT Policy and Innovation in Low-Tech*. Oslo: STEP – Centre for Innovation Research.

Hodne, F., & Grytten, O. H. (2000). *Norsk økonomi i det nittende århundre*. Bergen: Fagbokforlaget.

Karl, T. L. (1997). *The Paradox of Plenty*. Berkeley: University of California Press.

Kuznets, S. (1965). *Economic Growth and Structure*. New York: W. W. Norton & Company.

Landau, R., & Rosenberg, N. (Eds.). (1986). *The Positive Sum Strategy: Harnessing Technology for Economic Growth*. Washington, DC: National Academy Press.

Lange, E. (Ed.). (1989). *Teknologi i virksomhet: verkstedindustri i Norge etter 1840*. Oslo: Ad notam forlag.

Larsen, E. R. (2003). *Are Rich Countries Immune to the Resource Curse? Evidence from Norway's Management of its Oil Riches*. Discussion Paper 362 Statistics Norway.

Lederman, D., & Maloney, W. (2007). *Natural Resources: Neither Curse nor Destiny*. Washington, DC/Palo Alto: World Bank/Stanford University Press.

Locke, R. R. (1984). *The End of the Practical Man*. London: JAI Press Inc.

Lundgreen, P. (1990). Engineering Education in Europe and the U.S.A, 1750–1930: The Rise to Dominance of School Culture and the Engineering Professions. *Annals of Science, 47*(1), 33.

Lundvall, B.-Å. (1992). *National Systems of Innovation*. London: Pinter Publishers.

Maloney, W. F. (2002). *Missed Opportunities: Innovation and Resource-Based Growth in Latin America*. World Bank Policy Research Working Paper 2935.

Martin, W., & Mitra, D. (2001). Productivity Growth and Convergence in Agriculture and Manufacturing. *Economic Development and Cultural Change, 49*(2), 403–422.

Mehlum, H., Moene, K., & Torvik, R. (2005). Institutions and the Resource Curse. *The Economic Journal, 116*(508), 1–20.

Mendoza, R. U., et al. (2012, April 25). *Devil's Excrement or Manna from Heaven? A Survey of Strategies in Natural Resource Wealth Management*. Asian Institute of Management Working Paper.

Milward, A., & Saul, S. B. (1973). *The Economic Development of Continental Europe, 1780–1870*. London: Allen & Unwin.
Milward, A., & Saul, S. B. (1977). *The Development of the Economies of Continental Europe 1850–1914*. London: Allen & Unwin.
Mokyr, J. (2005). *Gifts of Athena*. Princeton/Oxford: Princeton University Press.
Mokyr, J. (2006). Knowledge, Enlightenment, and the Industrial Revolution: Reflections on Gifts of Athena. *History of Science, 45*(part 2), 185.
Mokyr, J. (2009a). Intellectual Property Rights, the Industrial Revolution, and the Beginnings of Modern Economic Growth. *American Economic Review, 99*(2), 349.
Mokyr, J. (2009b). *The Enlightened Economy. An Economic History of Britain 1700–1850*. New Haven: Yale University Press.
Musgrave, P. (1967). *Technical Change the Labour Force and Education*. Oxford/New York: Pergamon Press.
Nelson, R. R. (1968). Innovation. In *International Encyclopedia of Social Sciences*, 7.
Nelson, R. R. (1993). *National Innovation Systems a Comparative Analysis*. New York: Oxford University Press.
Nelson, B., & Behar, A. (2008, February 7). *Natural Resources, Growth and Spatially-Based Development: A View of the Literature*. World Development Report, Reshaping Economic Geography.
Nelson, R. R., & Winter, S. G. (1982). *An Evolutionary Theory of Economic Change*. Cambridge: Belknap Press.
North, D. (1955). Location Theory and Regional Economic Growth. *Journal of Political Economy, 63*, 243.
North, D. C. (1991). Institutions. *The Journal of Economic Perspectives, 5*(1), 97.
North, D. C. (1993). *New Institutional Economics and Development*. Washington, DC: Washington University.
North, D. C. (1996). Economic Performance Through Time: The Limits to Knowledge. EconWPA, Economic History, 9612004.
North, D. (2006). *A Conceptual Framework for Interpreting Recorded Human History*. George Mason University, Working Paper 75.
North, D. (2009a). *Violence and Social Orders*. The Annual Proceedings of the Wealth and Well-Being of Nations, Beloit.
North, D. (2009b). *Violence and Social Orders*. Cambridge: Cambridge University Press.
Nuur, C., & Laestadius, S. (2009). *Anomalies in the Resource Curse Paradigm: The Case of Sweden*. Trieta Research Report, The Royal Institute of Technology, KTH.

O'Brien, P. (2011). *Stages in the Evolution of a Western Regime for the Discovery, Development and Diffusion of Useful and Reliable Knowledge.* URKEW Discussion Paper No. 7, London School of Economics.

O'Neal, L., & Williamson, J. G. (Eds.). (2014). *Cambridge History of Capitalism* (Vol. II). Cambridge: Cambridge University Press.

O'Rourke, K. H., & Williamson, J. G. (1995). Education, Globalization and Catch-Up: Scandinavia in the Swedish Mirror. *Scandinavian Economic History Review, 43*(3), 287.

Odagiri, H., et al. (2010). *Intellectual Property Rights, Development and Catch-Up an International Comparative Study.* Oxford: Oxford University Press.

Oxford Encyclopedia of Economic History, Vol. II (2003): Cooperative Agriculture and Farmer Cooperatives. Oxford: Oxford University Press.

Petrovsky, N. (2004). *Does Natural Resource Wealth Spoil and Corrupt Governments? A New Test of the Resource Curse Thesis.* University of North Texas.

Polanyi, M. (1983). *The Tacit Dimension.* Gloucester: Peter Smith.

Prebisch, R. (1950). *The Economic Development of Latin America and Its Principal Problems.* New York: United Nations.

Roderick, G. W., & Stephens, M. D. (1972). *Scientific and Technical Education in Nineteenth-Century England.* Newton Abbot: David & Charles.

Sachs, J. D., & Warner, A. M. (2001). The Curse of Natural Resources. *European Economic Review, 45*, 4–6.

Sæther, B., Isaksen, A., & Karlsen, A. (2011, June). Innovation by Co-evolution in Natural Resource Industries: The Norwegian Experience. *Geoforum, 42*(3), 373.

Sandberg, P. (1979). The Case of the Impoverished Sophisticate: Human Capital and Swedish Economic Growth Before World War 1. *Journal of Economic History, 49*, 1.

Sandberg, L. (1982). Ignorance, Poverty and Economic Backwardness in the Early Stages of European Industrialization: Variations on Alexander Gerschenkron's Grand Theme. *Journal of European Economic History, 11*, 675–697.

Sanderson, M. (1994). *The Missing Stratum Technical School Education in England 1900–1990s.* London: Athlone Press.

Sejersted, F. (1993). *Demokratisk kapitalisme.* Oslo: Universitetsforlaget.

Smart, M. (2009). Natural Resource Diversity and Democracy. *The Economic Society of Australia, 28*, 366.

Spence, M. (1973). Job Market Signaling. *The Quarterly Journal of Economics, 87*(3), 355.

Statistisk sentralbyrå (1866–1940): Norges offisielle statistikk, *Norges Bergverksdrift.* Oslo: Statistisk sentralbyrå.

Statistisk Sentralbyrå. (1952). *Norges Offisielle Statistikk, Nasjonalregnskap 1930–1939 og 1946–1951*. Oslo: Statistisk Sentralbyrå.
Statistisk Sentralbyrå. (1953). *Norges Officielle Statistikk, Nasjonalregnskap 1900–1929*. Oslo: Statistisk Sentralbyrå.
Suslova, E., & Volchkova, N. (2007). *Human Capital, Industrial Growth and Resource Curse*. State University Higher School of Economics: WP 13/2007/11 CAS Working Paper Series.
The Dutch Disease. (1977, November 26). *The Economist*.
UNCTAD. (2001). *World Investment Report 2001 Promoting Linkages*. New York/Geneva: United Nations.
Van der Ploeg, F. (2010). *Natural Resources: Curse or Blessing?* CESifo Working Paper No. 3125.
Vernengo, M. (2006). Technology, Finance, and Dependency: Latin American Radical Political Economy in Retrospect. *Review of Radical Political Economics, 38*(4), 551.
Ville, S., & Wicken, O. (2013, October 1). The Dynamics of Resource-Based Economic Development: Evidence from Australia and Norway. *Industrial and Corporate Change, 22*(5), 1341.
Viner, J. (1952, February). International Trade and Economic Development. *Louvain Economic Review, 20*(1).
Wicken, O. (2010). *The Norwegian Path Creating and Building Enabling Sectors*. Working Paper, Centre for Technology, Innovation and Culture, University of Oslo.
Wrigley, E. A. (2010). *Energy and the Industrial Revolution*. Cambridge: Cambridge University Press.

Web Pages

UNDP. (2018, March 30). *Income Gini Coefficient*. Retrieved from http://hdr.undp.org/en/content/income-gini-coefficient
United Nations Development Programme. (2018, March 30). *Human Development Reports*. http://hdr.undp.org/en/data
World Data Bank. (2018, March 30). *World Development Indicators (WDI) & Global Development Finance (GDF)*. Retrieved from http://data.worldbank.org/data-catalog/world-development-indicators

2

An Innovative and Growing Mining Sector

Productions and Exports

The mining sector was small in terms of share of GDP. Numbers lack for the nineteenth century, but in 1910, mining and mineral and metal extraction stood for 1.2 per cent of GDP. In 1930, the sector stood for 0.9 per cent; in 1935 it constituted 0.9 per cent; and in 1939, 1.1 per cent.[1] The mining sector used a small share of total workers. In 1874, the sector used 0.5 per cent of the occupational population; in 1890 it used 0.6 per cent; and in 1920, 0.9 per cent.[2]

Yet, although the mining sector represented a small part of the Norwegian economy, it grew and expanded considerably. During the nineteenth century, the demand for metals and minerals increased due to the ongoing industrialisation processes that were taking place in Europe and America at the time. Norway took advantage of these changes and exported most of the metals, minerals, and alloys it produced. It was the expanding markets in other countries which gave

[1] Statistisk sentralbyrå, *National accounts 1900–1929* (Oslo, 1953); Statistisk sentralbyrå, *Nasjonalregnskap 1865–1960* (Oslo, 1965).

[2] Based on calculations of statistics from Statistisk sentralbyrå, *Historical statistics 1968* (Oslo, 1969).

K. Ranestad, *Knowledge-Based Growth in Natural Resource Intensive Economies*, Palgrave Studies in Economic History, https://doi.org/10.1007/978-3-319-96412-6_2

Norway the incentive to search for more mineral and metal ores and to increase production. According to official statistics most of the mineral and metal products were exported to Western European countries and the United States. Copper was mostly sent to Denmark, Germany, Holland, the United Kingdom, and the United States. Silver was mostly sent to Germany and iron was sent to Germany, the United Kingdom, and the United States. Pyrite was sent, for the most part, to Denmark, Germany, and the United Kingdom. Coal and nickel were exported to Sweden, Denmark, and Russia; and aluminium, which was produced from the early twentieth century, was distributed all over Europe and the United States.[3] The sector as a whole increased its share of exports from 6.5 per cent in 1900 to 29.2 per cent in 1939.[4]

This export-led sector depended largely on world prices of metals and minerals which were highly volatile. In the late 1870s, and then in the 1920s after the First World War, prices of multiple metals and minerals declined and led to recession periods for metal and mineral extraction industries in many different parts of the world. Prices of metals and minerals were considerably high during the First World War, which made the abrupt price decline from around 1922 challenging for most metal and mineral industries in Norway. Figure 2.1 illustrates the volatility of metal and mineral prices in the nineteenth and early twentieth centuries, and the severe downturns of the 1870s and the 1920s.

Mining in Norway has a long tradition and became an important economic sector in the sixteenth century. Traditional metals, such as copper, silver, gold, and iron, became key productions.[5] Central mining companies developed from the seventeenth century, notably the state-owned company Kongsberg Silver Works, established in 1623; Røros Copper Works, founded in 1644; and Løkken Works, established in 1654. In the nineteenth century, traditional metal productions, notably iron, copper, silver, and nickel, increased. In the 1860s, annual copper production was on average 14,000–15,000 tons and iron production was on average

[3] Norges Offisielle Statistik, *Norges Handel* (Oslo, 1860–1940).

[4] Statistisk sentralbyrå, *Nasjonalregnskap 1900–1929*; Statistisk sentralbyrå, *Nasjonalregnskap 1930–1939 and 1946–1951.*

[5] H. Carstens, *...Bygger i Berge* (Trondheim, 2000), p. 32.

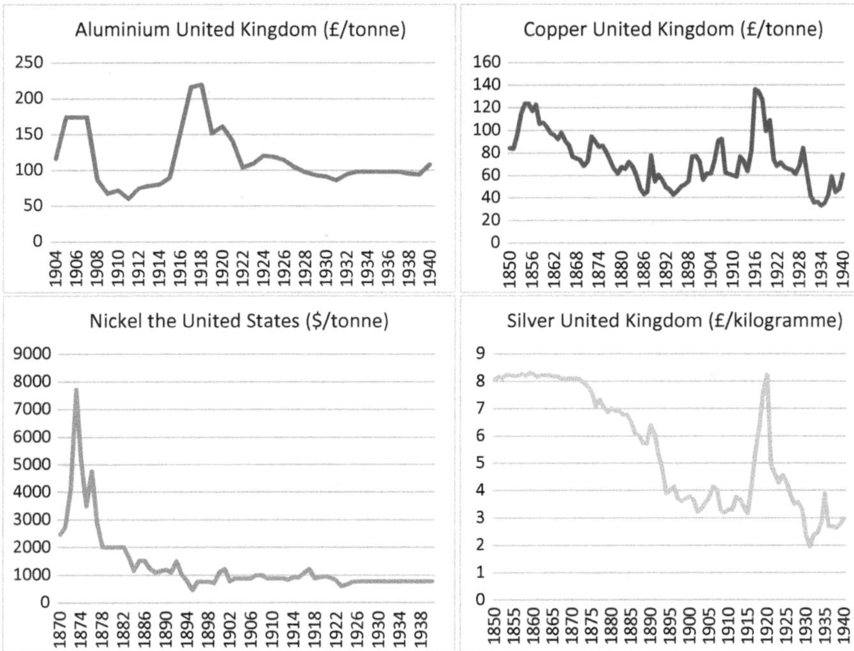

Fig. 2.1 Prices of selected metals. Source: C. J. Schmitz, *World Non-Ferrous Metal Production and Prices, 1700–1976* (Frank Cass 1979)

20,000–21,000 tons. In the 1870s, several mining productions suffered. Norway produced one-third of the world's total supply of nickel between 1873 and 1876—over 42,000 tons in 1876—but many of the nickel mines closed down in the late 1870s.[6] After the recession period in the 1870s, production of mining goods increased considerably. Copper production increased from around 10,000 tons in 1880 to 47,000 tons in 1900 and 50,000 tons in 1918; silver production increased from 885 tons in 1880 to 1600 tons in 1905 and 14,600 tons in 1930.

Initiatives were also taken to start up new metal and mineral productions. The diversification of the mining sector is shown in Table 2.1. First, pyrite was extracted from the mid-nineteenth century and became an important mineral production by the late nineteenth century. In 1870,

[6] Wicken, "The Norwegian Path Creating and Building Enabling Sectors", working paper, Centre for Technology, Innovation and Culture, University of Oslo, 2010, p. 15.

Table 2.1 Norwegian mining products in percentage (value: NOK)

	Percentage	
Mining products[a]	1919	1939
Mining ore products		
Silver ore	***6***	***0.2***
Copper ore	1.5	2.2
Pyrite (partly with copper)	***76.5***	***7.9***
Nickel ore	0.8	0.5
Iron ore	8.6	10.3
Zinc and lead ore	0.03	0.2
Molybdenite	4.7	1.1
Rutile	0.05	0.04
Chrome ore	0.2	–
Manganese ore	0.01	–
Titaniferous	0.9	–
Arsenopyrite	0.009	–
Apatite	0.5	–
Fluorspar	0.1	–
Other products	–	0.04
Electro-metallurgical products		
Silver	–	0.2
Copper	–	3.9
Nickel	**–**	***11.6***
Aluminium	**–**	***19.3***
Ferroalloys	**–**	***19.4***
Pig iron	–	2.5
Zink	–	6.5
Lead and tin	–	0.4
Sulphur	–	13.6

[a]The value of coal is excluded from the statistics
Calculations based on statistics from Norges offisielle statistikk, *Norges bergverksdrift 1919 and 1920*; Statistisk sentralbyrå, *Norges bergverksdrift 1939*

production of pyrite reached 67,000 tons, and went up to 162,000 tons in 1905 and more than 700,000 tons in 1930.[7] Pyrite production increased significantly in the early twentieth century, making it a key large-scale industry. In 1919 pyrite, together with copper, stood for almost 80 per cent of total mining production.

Second, production of coal was initiated in Svalbard in the early 1900s. In 1912, coal production was almost 33,000 tons and it reached

[7] Norges offisielle statistikk, *Norges Bergverksdrift* (Oslo, 1866–1940).

450,000 tons in 1924. Subsequently, production declined to 195,500 tons in 1931, but increased in the mid-1930s. In 1939, coal production totalled 611,700 tons.[8] Third, increasing global markets were accompanied by the branching out of a large-scale electro-metallurgical industry at the turn of the century. New large-scale electro-metallurgical productions emerged notably of aluminium, iron, nickel, and others, based on the utilisation of hydroelectric power, of which Norway had plenty. With large companies employing hundreds, and sometimes thousands of workers, it was found that "more than anything else it was this industry which transformed Norway from a rather poor farmland into a prosperous industrialised country".[9] Nickel production increased from 1900 tons in 1900 to more than 24,000 tons in 1918. Iron production increased from 6700 tons in 1880 to 18,000 tons in 1900 and more than 700,000 tons in 1915. After the recession period in the 1920s, production once again increased. Iron production totalled around 1,400,000 tons in 1939. Nickel production increased from around 2100 tons in 1922 to more than 31,000 tons in 1939. Aluminium production increased from 13,000 tons in 1923 to more than 31,000 tons in 1939.[10]

Thus, the general tendency was that metal and mineral productions increased, especially from the turn of the century, except for a downturn in the 1920s (see changes in production in Appendix 1). The number of workers employed in the mining industry increased, except for a couple of years in the 1920s, when the number decreased sharply. Before 1900, the average number of workers was around 2650, while between 1901 and 1920 workers totalled 6580 and between 1921 and 1939 the average number was around 8800.[11]

In addition, official mining statistics allow us to calculate how much the average worker produced of mineral and metal ore (removal of ore from mines) and pure metals, minerals, and alloys (smelting and refinement of ore at smelting plants) each year.[12] Appendix 2 shows production of

[8] Norges offisielle statistikk, *Norges Bergverksdrift* (Oslo, 1866–1940).

[9] H. Carstens, ...*Bygger i Berge* (Trondheim, 2000), p. 32.

[10] Norges offisielle statistikk, *Norges Bergverksdrift* (Oslo, 1866–1940).

[11] Norges offisielle statistikk, *Norges Bergverksdrift* (Oslo, 1866–1940).

[12] Detailed information about production and workers on a company level has not been obtained, except two.

selected metal and mineral ores per average worker and Appendix 3 shows production of selected metallurgical products per worker. Some of the productions showed great variations, notably coal, nickel, and iron, but the general tendency, both for mineral and metal ores and metallurgical products, is that the amount of metals and minerals that were produced annually per worker increased over time. There were variances within the sector, but in general terms, much larger quantities of metals and minerals were produced by each worker after the turn of the century than in the 1870s. Productivity seemed to increase radically in the whole sector.

A long-term analysis of two companies illustrates the dramatic increase in productivity from the turn of the century on a company level. The state-owned company Kongsberg Silver Works increased the average productivity from 1.3–3.4 tons of silver per worker during 1866–1903 to more than 5.8 tons in 1904, more than 34 tons in 1915, and more than 72 tons in 1932. Orkla Mining Company, a Swedish company which took over the operation of Løkken Works in 1904, produced less than 30 tons of pyrite per worker during 1880–1909, but this increased to 82 tons in 1910, to around 700 tons in 1926, and to more than 800 tons in the 1930s.[13]

The increase in production and exports, the diversification of the sector, and the increase in productivity were based on the adoption of new ways of finding, removing, and extracting ores. From the mid-nineteenth century, companies adopted new energy sources, notably mechanical power, and later electric power, new mining techniques, organisation and coordination systems, and new equipment, furnaces, converters, and smelting methods, which rationalised operations and enabled large-scale production. The next section of this chapter outlines these technological changes.

A Technological Transformation: An Outline

Increased production and production of new metals and minerals from the late nineteenth century led to the exhaustion of easy accessible high-grade mineral and metal ore deposits in many parts of the world. The

[13] Norges offisielle statistikk, *Norges Bergverksdrift* (Oslo, 1866–1940).

Table 2.2 Simple overview of challenges and changes in world mining from the late nineteenth century

Activity	New challenges	Technological solutions
Finding ores	Less available and low-grade mineral deposits	Systematic and organised ore surveys with the use of techniques to identify and detect specific metals, minerals, and mineral compositions
Removal of ores	Deeper, larger, and solid mines for large-scale production without collapsing	More precise surveys and mine measurement with the use of mechanised and electric equipment
Processing ores	Making pure products of ores less than 4 per cent	New concentration, separation, and refining techniques

Based on a review of C. Singer et al. (eds.) *A History of Technology, volume V The Late Nineteenth Century 1850 to 1900* (Oxford, 1958); Carstens, *...Bygger i Berge: en beretning om norsk bergverksdrift* (Trondheim, 2000)

increased demand, and exhaustion of mines, meant that companies met with new difficulties in terms of finding, removing, and processing ores and attention was given to new mineral- and low-grade ores in more isolated places and deeper underground. In order to maintain profitability in these new global settings, companies adopted new and more efficient techniques for finding, prospecting, and removing ores, and for organising and coordinating work to process the ores. Undoubtedly, mining became more challenging over time (see Table 2.2 for challenges and changes).

Before the eighteenth century, miners had little knowledge of mineralogical features and geological formations. Ore was found more or less randomly.[14] Throughout the eighteenth century, new minerals were recognised, described, and classified, and resulted in the discovery of many new elements, such as cobalt, nickel, manganese, tungsten, molybdenum, uranium, and others, and new and more efficient techniques to find ores developed. Systematic efforts led to more ore discoveries.[15] In Norway—as

[14] F. Habashi, *Schools of Mines* (Quebec, 2003), p. 61.

[15] J. A. S. Ritson, "Metal and Coal Mining, 1750–1875", in *A History of Technology, volume IV The Industrial Revolution The Industrial Revolution 1750 to 1850,* C. Singer et al. (eds) (Oxford, 1958), pp. 66–67.

in other countries—geological surveys, ore surveys, and mine measurements were gradually carried out in more systematic and organised ways partly due to the use of new methods. Chemical analyses, microscopes, and separation processes were increasingly used in ore analyses in the nineteenth century. The use of chemical processes made it possible to determine the composition of rocks and minerals, and, to a greater extent than earlier, to reveal their inner structure. This development was accompanied by an increased use of equipment and laboratories in geological research. The NGS of Norway used optical mineralogy, for example, from the 1870s, which involved thin pieces of mineral or rock placed under polarisation microscopes.[16] Horses were traditionally used to rotate drills but were later replaced by machines to make deeper holes. Improved geophysical methods were developed, and surface boring tools, such as augers and hand bores, were used to analyse the mineral deposits, their size, angles, directions, concentration, and so on. From the mid-nineteenth century, industrial diamonds for rotary drilling and prospecting were adopted.[17] From the 1920s, magnetic and electrical methods became common. By 1935, new methods consisting of systematic mapping of ore blocks were used.[18]

Constructing mines and removing ores were traditionally based on animal and human power and most of the rock underground was extracted manually with picks, crowbars, and wedges. Men and animals transported the ore by carrying the load on their backs and by pushing or pulling a wheeled tram.[19] In the latter half of the nineteenth century, the mining sector in Norway, as in several other countries, went through a mechanisation process. More powerful machinery and power sources became common, and mechanical and electric power enabled deeper and bigger mines and largely replaced steam, animal, and manual power. Machines of different types were used in the mines, notably steam engines, and later turbines,

[16] A. K. Børresen and A. Wale, *Kartleggerne* (Trondheim, 2008), pp. 29 and 74.

[17] J. A. S. Ritson, "Metal and Coal Mining, 1750–1875", in *A History of Technology, volume IV The Industrial Revolution The Industrial Revolution 1750 to 1850,* C. Singer et al. (eds) (Oxford, 1958), p. 68.

[18] A. K. Børresen and A. Wale, *Kartleggerne* (Trondheim, 2008), pp. 131–132.

[19] J. A. S. Ritson, "Metal and Coal Mining, 1750–1875", in *A History of Technology, volume IV The Industrial Revolution The Industrial Revolution 1750 to 1850,* C. Singer et al. (eds) (Oxford, 1958), p. 77.

which gave power to lifts, transport vehicles, and drainage equipment, and shaft-sinking and vertical excavations from the surface were made in deeper mines.[20] In 1895, 26 out of 90 companies used engines, with a total of 31 water engines, 32 steam engines, 18 electrical engines, and 1 gas engine.[21]

The adoption of electric power revolutionised world mining and also contributed to a rationalisation of mining operations in Norway. The first big electric power plant for mining was installed at Røros Copper Works in 1896 and included a high-voltage network which provided power to lifts, pumps, locomotives, crushing machines, and other equipment.[22] By the early twentieth century, power stations had become common, and eventually all companies used mechanical and electric power. In 1918, a total of 1654 electric engines and mechanical power worth 36,947 horsepower were spread around 169 companies.[23] In 1938, 146 companies used a total of 6592 electric engines. Moreover, the total primary power used in ore, metal, and mineral extraction and processing companies was 98,078 horsepower, and electricity for smelting, electrolysis, and so on totalled 294,210 kW (around 400,000 horsepower).[24]

New processing techniques enabled production of new metals, minerals, and alloys and rationalised ore extraction. Newly developed converters, furnaces, and crushing-, separation, and smelting techniques were developed and permitted more efficient utilisation of metal and mineral ores, and lower-grade ore. In Norway, a variety of separation and smelting techniques were experimented with, and more techniques were used from the late nineteenth century. The Manhés process was common in copper production from the late nineteenth century and was adopted in numerous copper companies.[25] The flotation technique spread rapidly and

[20] J. A. S. Ritson, "Metal and Coal Mining, 1750–1875", in *A History of Technology, volume IV The Industrial Revolution The Industrial Revolution 1750 to 1850,* C. Singer et al. (eds) (Oxford, 1958), 1958, p. 69.

[21] Norges Officielle Statistik, *Rigsforsikringsanstaltens Industristatistik for årene 1895–1899* (Kristiania, 1904).

[22] B. I. Berg, *Gruveteknikk ved Kongsberg Sølvverk 1623–1914* (Trondheim, 1998), p. 421; G. B. Nissen, *Røros Kobberverk 1644–1974* (Trondheim, 1976), pp. 195–96.

[23] There were 169 mining companies in Norway in 1917: Norges Officielle Statistik, *Industristatistikk for året 1918* (Kristiania, 1921).

[24] Norges Offisielle Statistikk, *Norges Industri Produksjonsstatistikk 1939* (Kristiania, 1941).

[25] J. H. L. Vogt, *Kobberets historie* (Kristiania, 1895).

eventually dominated ore processing. By the early twentieth century it was used on multiple low-grade ores, notably copper.[26]

The introduction of new energy sources and techniques at Røros Copper Works in the 1870s, 1880s, and 1920s illustrates the experimentation and non-stop adoption of new processing methods by mining companies during these periods. A drop in copper prices in the 1870s–1980s led Røros to make changes to save operational costs. One way to reduce expenses was to use other types of fuel. Members of the Board of Directors discussed the possibility of using coke instead of charcoal in the smelting process because the use of coke was much cheaper and easier to obtain. The company started to use coke, first mixed with charcoal, but later with more coke than charcoal and finally with coke alone.[27] Another way to reduce cost was to change the whole ore extraction system. A new special shaft furnace was installed with slag pots, which allowed for higher quantities of ore to be smelted with less coke and charcoal.[28] The next step was to upgrade and improve the quality of copper, and the possibility of replacing the current refining of copper with another process "applicable to mechanical processing" was considered.[29] In February 1887, after a long evaluation process, the company decided to adopt the Manhés process.[30] The process was put to use in October 1888 and it gave 99 per cent copper and "worked excellently", according to one of the Board members.[31] Around 15 years later, a Committee was organised to evaluate Røros Copper Work and the use of this technique was pointed out as particularly cost-effective. Around 150 NOK per ton copper was saved with the Manhés process, according to the Committee.[32] Copper smelting became more efficient and severe economic problems were solved with these changes.

[26] W. H. Dennis, *Metallurgy 1863–1963* (London, 1963), p. 28.

[27] J. P. Friis, *Direktør Jacob Pavel Friis' erindringer* (Oslo, 1944), pp. 173–174.

[28] G. B. Nissen, *Røros Kobberverk 1644–1974* (Trondheim, 1976), p. 191.

[29] Røros kobberverk, *årlig rapport 1886* (Trondheim, 1887), p. 8.

[30] Røros kobberverk, *årlig rapport 1887* (Trondheim, 1888), p. 8.

[31] Bergingeniørforening, *Tidsskrift for bergvæsen* (Oslo, 1916), p. 43.

[32] *Beretning angående Røros Verks tilstand fra den i generalforsamling i verket den 29. april 1902 Nedsatte komité* (Trondhjem, 1902), pp. 49–50.

In the early 1920s, Røros faced new economic problems. The new challenge involved the utilisation of low-grade ores, which was a common problem in mining. A significant portion of the known ore had been removed from Storvarts Mine, and only low-grade pyrite ore remained. It was then discussed whether a more modern ore-dressing technique could be used. Flotation—which separated minerals from the gangue in a liquid pulp by means of air bubbles and allowed for processing of ore containing 1.5 per cent copper—was adopted in Røros in 1927 and 1932.[33] Professor Pedersen explained in 1929 that "ore that was considered to be worthless before" was now being utilised.[34]

The amalgamation process and the use of sodium cyanide were used to produce silver. From the turn of the century, electric smelting and electrolysis were in production of aluminium, nickel, and steel, and magnetic separation was used in iron production.[35] Experimentation and adoption of new processing techniques took place on a wide scale in Norway and permitted the extraction of lower-grade ores and the making of purer products.

Summary

Most of the metals and minerals that were produced in Norway were exported to European industrial powers and American countries, and were dependent on, and vulnerable to, changes in global trends and prices. In the 1870s, and then in the 1920s, when prices on minerals and metals declined dramatically, production also fell. However, the general tendency was that the global use and demand of metals and minerals for industrial purposes increased and Norway had a large market abroad for its metal and mineral products. Mineral and metal goods traditionally ranged from copper, nickel, and iron to silver, but more products were

[33] H. Pedersen, «Samarbeidet mellem teknisk-videnskapelig forskning og industrien, foredrag i Polyteknisk Forening den 29. oktober 1929, *Teknisk Ukeblad* (2 January 1930), p. 6.

[34] H. Pedersen, «Den tekniske høiskole er i kontakt med storindustrien», *Bergens Tidende* (9 April 1929).

[35] A. K. Børresen and A. Wale (red.), *Vitenskap og teknologi for samfunnet? Bergfagene som kunnskapsfelt* (Trondheim, 2005).

made, production of pyrite increased radically, and the sector diversified and branched out a large-scale electro-metallurgical industry producing aluminium, iron alloys, molybdenum, and so on by the turn of the century. Production and exports increased dramatically from the turn of the century and productivity seemed to increase drastically across the whole sector.

The mining industry worldwide met with new challenges in terms of finding, removing, and processing ore, which led to radical changes in operational techniques. In Norway, new power sources were adopted on a wide scale during the nineteenth century, and replaced steam, animal, and manual power. Increasingly complex mining constructions, and new machines and diverse crushing, separation, and smelting methods were installed and new ore dressing and smelting techniques permitted the utilisation of lower-grade ores. This radical technological transformation permitted the new productions, the diversification, and the radical production and productivity increase in mining in Norway during this period.

Mining companies in Norway were characterised as being technologically up-to-date. Orkla Mining Company and Sulitjelma Mining Company were technologically advanced corporations with large-scale productions of copper and pyrite. By the 1920s, Orkla Mining Company was viewed as "an extremely well managed enterprise, the mine was one of the best in Europe and the mechanical equipment was excellent".[36] The company was called "Europe's most modern and well-equipped and leading pyrite works".[37] Sulitjelma Mining Company became Norway's second largest company at the beginning of the twentieth century. With the first railway in northern Norway and one of the first electric smelting ovens, the company was considered "Europe's most modern mining works".[38] The state-owned company Kongsberg Silver Mines became a "fully modern mining works" in 1914 with the installation of fully electric operation of pumps and lifts.[39] Coal companies also used electric

[36] T. Bergh et al. *Brytningstider: Orklas historie 1654–2004* (Oslo, 2004), p. 57.

[37] Bergingeniørforening, *Tidsskrift for kemi og bergvæsen* (Oslo, 1933), p. 49.

[38] A. K. Børresen and A. Wale, *Kartleggerne* (Trondheim, 2008), p. 107.

[39] O. A. Helleberg, *Kongsberg sølvverk 1623–1958: kongenes øyesten – rikenes pryd* (Kongsberg, 2000), p. 292; B. I. Berg, *Gruveteknikk ved Kongsberg Sølvverk 1623–1914* (Trondheim, 1998), p. 421.

power and multiple electric machines, but they used different types of extraction methods, depending on the hardness of the mineral, long wall machines, and electric drilling machines in the mines. One method was the adoption of electric trolley locomotives, and a lift transported the coal to a "modern American storage- and cargo arrangement".[40] Electro-metallurgical companies producing iron, steel, aluminium, and other minerals and metals became one of the cornerstones of the Norwegian economy in the twentieth century.

What are the underlying factors explaining the developing and innovative mining sector? Multiple factors contribute to explaining the technological transformation that took place, for example capital, access to foreign markets, and so on, yet the rest of the book focuses on the knowledge on which the technological changes were based. It explores (1) the knowledge on which the technological transformation was based and how learning was transformed into innovation, and (2) the functions of key knowledge organisations involved in these processes. Based on what knowledge did mining develop in Norway? How can we explain the underlying learning and innovation processes? Which organisations were involved?

Bibliography

Secondary Sources (Literature)

Beretning angående Røros Verks tilstand fra den i generalforsamling i verket den 29. april 1902 Nedsatte komité. (1902). Trondhjem: Adresseavisens Bog- & Aksidenstrykkeri.

Berg, B. I. (1998). *Gruveteknikk ved Kongsberg Sølvverk 1623–1914*. Trondheim: Senter for teknologi og samfunn, NTNU.

Bergens Tidende. (1929).

Bergh, T., et al. (2004). *Brytningstider: Orklas historie 1654–2004*. Oslo: Orion forlag.

Bergingeniørforening. (1916). *Tidsskrift for bergvæsen*. Oslo.

Bergingeniørforening. (1932). *Tidsskrift for kemi og bergvæsen*. Oslo.

[40] Bergingeniørforening, *Tidsskrift for kemi og bergvæsen* (Oslo, 1932), p. 183.

Bergingeniørforening. (1933). *Tidsskrift for kemi og bergvæsen*. Oslo.
Børresen, A. K., &Wale, A. (red.). (2005). *Vitenskap og teknologi for samfunnet? Bergfagene som kunnskapsfelt*. Trondheim: Tapir akademisk forlag.
Børresen, A. K., & Wale, A. (2008). *Kartleggerne*. Trondheim: Tapir akademisk forlag.
Carstens, H. (2000). *...Bygger i Berge: en beretning om norsk bergverksdrift*. Trondheim: Norsk bergindustriforening Den norske bergingeniørforening Tapir.
Dennis, W. H. (1963). *Metallurgy 1863–1963*. London: Routledge.
Friis, J. P. (1944). *Direktør Jacob Pavel Friis' erindringer*. Oslo: Cammermeyers Boghandel.
Habashi, F. (2003). *Schools of Mines the Beginnings of Mining and Metallurgical Education*. Québec: Métallurgie Extractive Québec.
Helleberg, O. A. (2000). *Kongsberg sølvverk 1623–1958: kongenes øyesten – rikenes pryd*. Kongsberg: Forlag Langs Lågen i samarbeid med Sølvverkets venner.
Nissen, G. B. (1976). *Røros Kobberverk 1644–1974*. Norway: Aktietrykkeriet.
Norges Offisielle Statistik. (1860–1940). *Norges Handel*. Oslo: Statistisk sentralbyrå.
Norges offisielle statistikk. (1866–1940). *Norges Bergverksdrift*. Oslo: Statistisk sentralbyrå.
Norges Offisielle Statistikk. (1941). *Norges Industri Produksjonsstatistikk 1939*. Kristiania: H. Aschehoug & Co.
Rigsforsikringsanstalten. (1904). *Norges Officielle Statistik, Rigsforsikringsanstaltens Industristatistik for årene 1895–1899*. Kristiania: H. Aschehoug.
Røros kobberverk. (1886). *årlig rapport 1886*. Trondheim: Røros kobberverk.
Røros kobberverk. (1887). *årlig rapport 1887*. Trondheim: Røros kobberverk.
Schmitz, C. J. (1979). *World Non-Ferrous Metal Production and Prices, 1700–1976*. London: Frank Cass.
Singer, C., et al. (Eds.). (1958). *A History of Technology, volume IV The Industrial Revolution 1750 to 1850*. Oxford: The Claredon Press.
Statistisk Sentralbyrå. (1921). *Norges Officielle Statistik, Industristatistikk for året 1918*. Kristiania: H. Aschehoug & Co.
Statistisk Sentralbyrå. (1952). *Norges Offisielle Statistikk, Nasjonalregnskap 1930–1939 og 1946–1951*. Oslo: Statistisk Sentralbyrå.
Statistisk Sentralbyrå. (1953a). *Norges Officielle Statistikk, Nasjonalregnskap 1900–1929*. Oslo: Statistisk Sentralbyrå.
Statistisk Sentralbyrå. (1953b). *National accounts 1900–1929*. Oslo: H. Aschehoug.

Statistisk sentralbyrå. (1965). *Nasjonalregnskap 1865–1960*. Oslo: Statistisk sentralbyrå.
Statistisk sentralbyrå. (1969). *Historical statistics 1968*. Oslo: Statistisk sentralbyrå.
Teknisk Ukeblad. (1930). Kristiania: Den norske ingeniør- og arkitektforening og Den polytekniske forening.
Vogt, J. H. L. (1895). *Kobberets historie*. Kristiania: P. T. Mallings boghandels forlag.
Wicken, O. (2010). The Norwegian Path Creating and Building Enabling Sectors, Working Paper, Centre for Technology, Innovation and Culture, University of Oslo.

Part II

Knowledge Development in Technologically Complex Mining: A Framework

The aim of this part is to present a framework of the knowledge on which the innovative and developing mining sector in Norway was based. As mines became gradually deeper and bigger, more advanced mathematics and precise work and accuracy were used to prevent them from collapsing. New power sources, machinery, tools, and equipment of different types eased the physical work for miners and engineers but required insight into new knowledge areas. Technological changes were based on knowledge specialisations and complex learning and knowledge bases, which are here categorised into "scientific knowledge areas", notably mathematics, physics, mechanics, geology, mineralogy, chemistry, and metallurgy, and also "engineering", "management", and "economics", "tacit knowledge", "practical knowledge", and "know-who", "know-what", "know-how", and "know-why". This knowledge framework sets the starting point for the analysis of the activities and performances of knowledge organisations in Norway.

3

Catching Up with World Mining: A Model of Mining Knowledge

Scientific Knowledge Areas

The uniqueness of each ore deposit and mine makes each mining project unique. Yet, in the period studied here, some general common traits can be outlined related to the three main procedures which were required to produce pure metals and minerals, and to make alloys: (1) geological mapping and ore surveys, (2) removal of ore from the ground, and (3) ore processing. First, ore deposits had to be found and analysed. Second, the ore was dug out and removed from the earth. Third, the ore was transported to be processed, separated, smelted, and made into pure metals (see Table 3.1).

The knowledge and working tasks on which these operational activities were based changed a great deal over time. From the turn of the nineteenth century geological surveys, mapping, finding ores, and mine measurements were carried out in more systematic and organised ways, and with the adoption of new ore survey and prospecting methods and equipment, such activities became based on extensive knowledge in the field of geology, mineralogy, chemistry, and, subsequently, electro-engineering and magnetism.

K. Ranestad, *Knowledge-Based Growth in Natural Resource Intensive Economies*, Palgrave Studies in Economic History, https://doi.org/10.1007/978-3-319-96412-6_3

Table 3.1 Mining

Production branches	Operational activities				
				Processing of ore	
Metal ores Non-metallic minerals Stones or rocks Energy minerals	Geological mapping and ore surveys	Removal of ore	Crushing and milling	Concentration and separation techniques to remove gauge	Smelting and refining

Based on H. Carstens, ...*Bygger i Berge: en beretning om norsk bergverksdrift* (Trondheim, 2000)

An important part of mining was to plan operation and forecast economic profit. This work was based on extensive use of knowledge in mathematics, economics, and economic geology. Mining laws and commercial, civil, and social laws of the country were also vital in this work. An example of an increased focus in this area is the planning and forecasting that were implemented by Orkla, a Swedish company which took over the operation of Løkken Works in 1904. Before 1904, little knowledge of the extension of the ore deposits was collected and production depended on the extraction of the highest copper-containing ore. Old mining operations were arguably implemented "from hand to mouth". Mining engineer Holm Holmsen became technical director at the company and, in 1904, he made a report of the "ore in sight" and its percentage of copper and sulphur.[1] Røros Copper Works started earlier with such economic planning. In 1872, the mining engineer Jacob Pavel Friis presented a plan to the Directory of Røros Copper Works, which showed future production of two mines, Storwarts and Mugg, for respectively 70 and 30 years.[2] Detailed planning of future production became increasingly important from the late nineteenth century, especially as high-grade ore was running out and mines were exhausted.

[1] Orkla Grube-Aktiebolag, *Løkken Verk En Norsk Grube Gjennom 300 År* (Trondhjem, 1954), pp. 353–355.

[2] J. P. Friis, *Direktør Jacob Pavel Friis' erindringer* (Oslo, 1944), p. 169.

After analysing the ore and evaluating economic profit, one of the biggest challenges was to plan mine structures, that is map tunnels, adits, shafts, ventilation shafts, and so on, and to organise where they would go relative to each other. Making tunnels and adits meet below the ground surface required accuracy, precision, and detailed measuring. Constructions and structures beneath the earth were particularly difficult to make, since the mines could easily collapse.[3] Detailed and careful examinations of the geological ground were part of the preparation work, and physical and mathematical principles and methods in mechanics were key to designing and making mines. As mining developed, and mines became deeper and bigger, the construction of mines became increasingly challenging. Specialised knowledge in geometry, geology, mineralogy, mechanics, and construction engineering was essential and the mathematics used in operation became more advanced.

The mine preparation in the 1880s of the engineer and sous-director at Vigsnes Copper Works Emil Knudsen illustrates how mathematical principles and knowledge of natural sciences were combined with detailed measuring work and calculations. The old maps and observations at the firm were found to be inaccurate, and so Knudsen sought to make an accurate geological map of the oldest mine using a new theodolite (measuring device). After he had made systematic analyses of the mineral composition, he found out where the mineral ore developed underground:

> On the basis of my geological maps of mine, I did surveys and found vein number 1 again at 360 meters' depth, whereas vein number 3 in my theory was not to be found that deep since the cavity with ore formation had been removed from the transversal plateau. [...] [A]t 347 meters the ore appeared. [...] However, I found two new smaller ore veins at 260 and 286 meters [...] and when we got down to 460 meters I found vein number 2A, vein number 5 [...] and vein number 1. At 160 meters I found an unknown ore vein number 1B, which I also rediscovered at 360 meters.[4]

[3] J. A. S. Ritson, "Metal and Coal Mining, 1750–1875", in *A History of Technology, volume IV The Industrial Revolution The Industrial Revolution 1750 to 1850*, C. Singer et al. (eds) (Oxford, 1958), 69.

[4] F. Sæland, *Bergingeniør Emil Knudsens erindringer Inntrykk fra et bergmannsliv 1856–1897* (Kongsberg, 2005), 88–89.

His work led to "measurements [which] matched so accurately that there were only five millimetres difference in the lateral direction and 10 centimetres vertically". A "room and pillar" technique was used to remove the ore from the mine, in which the mined material was removed across a horizontal plane, creating horizontal arrays of rooms and pillars. Knudsen explained that figuring out the size of the pillars was crucial because too big pillars would potentially mean loss of ore and too small pillars could give too little support and involve the risk of the mine collapsing.[5] It is evident that such measurements and constructions were based on in-depth precision, and extensive knowledge of natural sciences, notably geology, mineralogy, as well as mathematics and physics.

New Knowledge Specialisations : Chemistry, Metallurgy, Mechanical Construction, and Electro-Engineering

The mechanisation process that started in the mid-nineteenth century, including the use of mechanised lifting, drainage, drilling, and transport equipment, gradually modified the daily tasks for miners and engineers. In 1926, the Norwegian mining engineer Wolmer Marlow stressed the importance of having knowledge of how to operate and repair new machinery:

> In order to master the increasingly applied mechanical operations at a modern mine and the often necessary large building projects, it must be required of the leader, that he fully knows this very important side of a mining operation. The daily and constant work at a mine is often purely mechanical and structural.[6]

[5] F. Sæland, *Bergingeniør Emil Knudsens erindringer Inntrykk fra et bergmannsliv 1856–1897* (Kongsberg, 2005), 8.

[6] Own translation: Bergingeniørforening, *Tidsskrift for kemi og bergvæsen* (Oslo, 1926), p. 31.

Eleven years later, in 1937, Professor Harald Dahl stressed that the increased use of mechanised equipment in mining strongly influenced the work of engineers. The faster the mechanisation, the more important it was for engineers to become familiar with the machinery. He referred to an English mining engineer who said that "the more machinery that gain access to the mines, the higher the requirements of expertise".[7] The installation, use, and repair of mechanical equipment and machinery were largely based on knowledge in mechanics, construction engineering, and related knowledge fields, which specialisations became key from the turn of the century when the use of mechanical power had become common.

Large power plants were used in mining from the late nineteenth century to provide electric power to equipment at mines and mineral- and metal-processing plants. The increased use of electric power was based on extensive knowledge in electro-engineering, which can be illustrated by the planning and installation of a power plant close to Røros in 1895, described in the mining engineer Emil Knudsen's memoirs. In the early 1890s, Knudsen explored the geological conditions at the Arv Lake, near the King's Mine, to find out whether it would be possible to install a power plant at this location. The first step of the process involved reading about electro-engineering.[8] The next step was to make multiple geological analyses and measurements, after which he found out that the waterfall at Arv Lake was too small for the mines at Røros, so he began considering waterfall conditions at other places. Thereafter he needed to learn how electric power plants functioned in practice and travelled to Norberg in Sweden, among other places, to see how companies used them there. Finally, after a long evaluation process, Knudsen considered the Kuraas waterfall in Glommen, 7–8 kilometres away from Røros, to be the best option, and in 1895 the installation plans were initiated.[9] The power station used three-phase electric power with transformers and cables between the mines and the primary station of around 4000 volts.

[7] Bergingeniørforening, *Tidsskrift for kemi og bergvæsen* (Oslo, 1937), p. 158.

[8] F. Sæland, *Bergingeniør Emil Knudsens erindringer Inntrykk fra et bergmannsliv 1856–1897* (Kongsberg, 2005), pp. 112–113.

[9] F. Sæland, *Bergingeniør Emil Knudsens erindringer Inntrykk fra et bergmannsliv 1856–1897* (Kongsberg, 2005), pp. 112–113, 116–117.

The waterfall provided "electric power" in the form of electricity to engines for transport, drilling, ore dressing, lifting, and lightening to Storvarts and King's Mines.[10] It was clear that both specialised knowledge in theoretical electro-engineering and practical understanding of how power stations worked were important.

After the ore was crushed and milled—often by using mechanised crushers and milling devices—the ore was processed through separation and smelting. The choice of separation, converting, and smelting techniques depended extensively on the characteristics and composition of the ore. The development of new techniques, and modifications of existing ones, was based on multiple knowledge specialisations within natural sciences and engineering. Testing, experimentation, creation, and use of an increased number of metal and mineral products—and constantly more, better, and more efficient techniques to process them—were built on in-depth knowledge of each mineral and metal and their characteristics and treatment. In 1925, the Norwegian mining engineer Fredrik Sebastian Nannestad confirmed that it was important to be aware that, in practice, there were no minerals which were treated equal. Therefore, the knowledge of ore-dressing methods, mineral characteristics, and processing techniques represented very extensive knowledge.[11] The broad knowledge base was also emphasised in Latin America. At a conference in Bolivia in 1928, the engineer Félix Cremer argued much in the same way as Nannestad. He underlined that the scope of mining was so vast, and the total knowledge that it involved was of such proportions that "no human brain [could] adequately meet its demands".[12] More generally, knowledge of ore features, and how ore was processed and made into pure metals and minerals, rested on extensive and specialised knowledge of geology, mechanical engineering, chemistry, metallurgy, and electro-metallurgy.

The making of the "Orkla process", which was a method for separating pure sulphur and copper from pyrite, illustrates the use of electro-engineering, combined with in-depth knowledge of mineralogy, geology,

[10] *Teknisk Ukeblad* (Kristiania, 2nd of mai 1895), p. 163.

[11] Bergingeniørforening, *Tidsskrift for kemi og bergvæsen* (Oslo, 1925), pp. 215–217.

[12] *Boletín de la Sociedad Nacional de Minería* (Santiago, 1928), p. 55.

and chemistry, in mineral processing. Instead of the very common flotation process, which was developed in the early twentieth century, Orkla Mining Company developed an alternative method which could be used on very fine pyrite, and which meant reduced crushing and grinding costs. The development of this process started in 1918, when chemist and metallurgist Harald Pedersen began exploring, in his laboratory, the possibility of utilising the sulphur content in pyrite without losing too much of its components. The experiments constituted of mixing crushed pyrite with coke and blowing air through the mixture after heating. The roasting and reduction process was carried out at a low temperature to avoid the liquid from agglomerating. Problems in the experimentation process led to the transfer of the whole research project to Portugal, and later to Oskarshamn Copper Works in Sweden, where further separation and smelting experiments were carried out. In 1927, the laboratory work was transferred to Løkken Mine with the necessary equipment, including a water-jacket furnace, electro filter for dust and sulphur, steam boiler, and other devices. At first, 60–75 per cent of the sulphur content of the pyrite was utilised. After further modifications—notably the installation of a catalyst chamber and an exhaust system to further benefit the sulphur content of the gas—up to 70 per cent, and later 80 per cent, of the mineral was utilised.[13] The development of this process, as well as the practical use, depended on theoretical knowledge of, as well as practical experience with, electro-engineering, geology, mineralogy, and chemistry.

The making of the Söderberg process, successfully used in aluminium production, is another example illustrating how processing techniques were based on, and depended on, numerous natural scientific knowledge areas as well as electro-engineering. The idea started with the Swedish-Norwegian Carl Wilhelm Söderberg, an electro-engineer from the Technical School in Hannover in Germany. He believed that it would be possible to develop a continuous, self-determining electrode for electric ovens, instead of the current one which needed to be replaced after five

[13] Orkla Grube-Aktiebolag, *Løkken Verk En Norsk Grube Gjennom 300 År* (Trondhjem, 1954), pp. 441–448.

to ten days.[14] Söderberg made experiments at Jøssingfjord Manufacturing Company in 1910, but his attempts were not successful. As Söderberg lost faith, the electro-chemist Mathias Sem, with a doctoral degree from the Technical School in Darmstadt, continued the work of Söderberg. After many tests, and many failed experiments, it was Jens Westly, a technician and chemist from the Technical School in Bergen, who finally solved the problem. He made a fastener, which was vital for the use of the electrode, and thus found the solution which made it possible to put into practice the continuous self-burning electrode.[15] The experimental process behind the Söderberg method illustrates how varied and long theoretical training and work experiences in numerous knowledge areas, and the combination of different skills, were used to create new processing techniques.

New Knowledge Specialisations in Mining Organisation: Management and Business Administration

Knowledge of administration and economics might seem unnecessary in the field of mining at first sight. Business historian Alfred D. Chandler wrote that administrative challenges in mining in the early twentieth century were minor compared to those of many other industries due to fewer technological changes.[16] However, this argument is unconvincing. On the contrary, it seemed to be highly important to know how to administrate and manage mining companies, in particular large-scale mining companies. The company structure varied from company to company, but mining companies were characterised by a strict division of labour, work hierarchy, and a strong organisation of workers from early on, and Fritz Hodne in fact points out that the mining sector in Norway, in the early

[14] A. K. Børresen and A. Wale (red.), *Vitenskap og teknologi for samfunnet? Bergfagene som kunnskapsfelt* (Trondheim, 2005), p. 15.

[15] K. Sogner, *Elkem gjennom 100 år Skaperkraft 1904–2004* (Oslo, 2003), pp. 59–62.

[16] A. Chandler, *Scale and Scope: the dynamics of industrial capitalism* (the United States, 1990).

nineteenth century, had stronger organisation than in other sectors.[17] Companies grew in size and had sometimes hundreds or thousands of workers. The largest mining companies in Norway in 1940 were Sulitjelma Mining Company with around 800 workers, Sydvaranger Ltd. with around 950 workers, and Sydvaranger Ltd. with 1441 workers. Other companies varied from a couple of workers to around 300 workers.[18] A larger workforce was more challenging to organise and added to the complexity of mining operations.

Mining companies normally had multiple divisions and the physical distance between mines and crushing and smelting plants—the different places where practical work occurred—meant that management systems of mining companies were complex, and that coordination systems needed to be set up across large geographical areas. The director, managers, and middle managers—positions largely meant to be held by formally trained mining engineers—were in charge of this work. These strategic technical, management, and middle management positions involved delegation of work, practical engineering tasks, designing mines, coordination work, technical decision-making, supervision, administration and coordination of several hundred workers, economic and strategic technical planning, accounting, scientific ore analyses, and maintenance of machinery and equipment.[19]

The head engineer, or "mine manager", planned and organised the work at the mine and was in charge of the equipment that was used and of the miners who removed the ore from the mines. This meant having the responsibility for the daily operation concerning removal and transport of ore, and planning and ensuring that the work was performed successfully. The mine manager normally had an assistant, or several, who helped him in his work and replaced him when he was away. Administrators of crushing plants were responsible for the workers and the equipment at the quarries, and supervisors of the processing and smelting plants

[17] F. Hodne and O. H. Grytten, *Norsk økonomi i det nittende århundre* (Bergen, 2000).

[18] Norges offisielle statistikk, *Norges Bergverksdrift* (Oslo, 1940).

[19] K. Hunstadbråten, *Blaafarveværket* (Drammen, 1997); F. Sæland, *Bergingeniør Emil Knudsens erindringer Inntrykk fra et bergmannsliv 1856–1897* (Kongsberg, 2005); S. Imsen and H. Winge, *Norsk historisk leksikon*, 2. edition (Oslo, 1999); *Lov, ang. Røraas Kobberværk*, 1818.

administered the ore smelting processes and were normally called "smelting masters", "smelting accountants", and metallurgists. They were in charge of ore testing and analysis and supervised the workers who operated the smelting ovens, converters, and furnaces. Bookkeeping and accounting were also tasks for the smelting accountant and the "mining accountant". "Managers" of mining companies normally administered the daily operations and delegated work to middle managers. "Directors" (or directories) had the overall responsibility and task of decision-making regarding the adoption of new techniques.[20] Increased production and large-scale companies from the late nineteenth century led to an increased focus on mining administration, long-term planning, management, and business organisation. In 1925, the *Norwegian Mining Journal* published an article saying that the understanding of "economic, commercial management of a mine" was important, and even required, for the employment in leading positions.[21] Similar arguments were found in other places around the world. The *Mining Bulletin* in Chile wrote that "[t]he administration of a mining business requires, in a larger degree than other engineering firms, these qualities of good management, without which the entire technique and mining experience are useless".[22] It is clear that knowledge about how to administrate ever-larger mining organisations efficiently became key.

Tacit and Practical Knowledge

The specific nature of mining made detailed and in-depth knowledge of each local geological setting, of the ground, and of ore deposit crucial. Geological structures, size of ore deposits, and the composition of mineral ore made work conditions at each mine, and extraction site, different. Open-pit mining was different from underground mining; coal was a softer material than, for example, copper and silver, which in turn had implications for the use of methods; horizontal and vertical mining

[20] S. Imsen and H. Winge, *Norsk historisk leksikon*, 2. edition (Oslo, 1999).

[21] Bergingeniørforening, *Tidsskrift for kemi og bergvæsen* (Oslo, 1925), p. 186.

[22] Own translation: *Boletín de la Sociedad Nacional de Minería* (Santiago, 1925), pp. 663–664.

required different types of structures and techniques, and minerals and metals reacted differently in chemical processes, which had implications for the selection of processing methods.

Figuring out which types of equipment and methods should be used in each mining location required practice and experimentation, trying out, modification, adaptation, and so on. The development of the Söderberg process (described in a previous section) exemplifies this trying and failing. Moreover, the daily tasks in mining were based on physical tasks of hands-on use of equipment and measurement tools, drawing and making of mines, removing, transporting, crushing and smelting ores. This sector—as well as other natural resource–intensive industries—has been found to be based to a large degree on learning by doing and trial and error, perhaps more so than in other industries.[23] In mining, in particular, this focus on practice and learning by doing is probably related to the large degree of tacit knowledge—or knowledge which is hard to explain—which seems to be present in this field.

How did miners and engineers acquire the skills, or the capacity, to carry out practical tasks? Much learning and capacity building happened by assessing and trying out different mining techniques and equipment at the working site. Learning how to select the proper technique for a mining plant, for example, and how to operate it successfully, required observing it in operation and trying it out in practice. One among many examples is the adoption of new drills at Røros Copper Works in the late nineteenth century, which involved a trip to the United States by mining engineer Emil Knudsen to learn about different drilling machines on the market. In the United States, Knudsen observed numerous drills "in action" and picked out one particular percussion-drilling machine called Marwin, which drilled 15 feet drill holes per hour in hard limestone. He recommended the Marwin drills to the Directory of Røros as he found them to be "the best" option for two of the company's mines.[24] Without the practical experience with different types of drills, detailed knowledge of their strengths and weaknesses, and without doing a practical evalua-

[23] H. Hirsch-Kreinsen et al., "Low-Tech Industries and the Knowledge Economy: State of the Art and Research Challenges", *PILOT Policy and Innovation in Low-Tech, STEP – Centre for Innovation Research* (Oslo, 2003), p. 9.

[24] *Teknisk Ukeblad* (Kristiania, 2 May 1895), p. 164.

tion of how the Marwin drills worked compared to other machines, it would have been hard, maybe even impossible, for Knudsen to select the best ones for the mines at Røros.

Know-What, Know-How, Know-Who, and Know-Why

Much experimentation and new mining and processing techniques had their origin in industrial powers, notably Britain, Germany, France, and the United States.[25] British and North American engineers, for example, experimented with new and more efficient mining machinery during the nineteenth century, such as drills and crushers. Hand-operated ore-dressing machines separated the metal from the barren rock, but John Taylor introduced a more efficient mechanical type in 1804 at the Wheal Crowndale mine in Devon. The first mechanical machine to crush the rock into smaller parts called "jaw breaker" was invented by Eli. W. Blake of New Haven in the mid-nineteenth century. The next advance in crushing was the gyratory crusher, introduced by P. W. Gates in 1881. This machine had more capacity than the previous jaw breaker. Multiple grinding machines were developed to pulverise the ore, of which some of the most common ones were the cylindrical mill and the ball mill.[26]

New concentration, separation, and smelting techniques were developed by metallurgists and chemists from Europe, Australia, and America to cope with lower-grade mineral and metal ores. R. Chadwick writes that

> [t]he second half of the nineteenth century was characterized by tremendous advances both in the method and scale of operation of metallurgical processes for the established metals, and in the invention and development of method[s] of extracting metals hitherto regarded as chemical curiosities.[27]

[25] See the general development of inventions and use of technology in mining in Singer et al. (eds.), *A History of Technology*, 5 volumes (Oxford, 1954–).

[26] W. H. Dennis, *Metallurgy 1863–1963* (London, 1963), pp. 22–27.

[27] R. Chadwick, "New Extraction Processes for Metals", in *A History of Technology, volume V The Late Nineteenth Century 1850 to 1900*, C. Singer et al. (eds.) (Oxford, 1958), p. 72.

The Manhés process, developed by the French Pierre Manhés in the 1880s for copper processing, was based on the English engineer Henry Bessemer's iron-processing technique, which in turn was based on the North American William Kelly's experimental work with iron.[28] This technique produced purer copper in much shorter time than other methods; it rationalised the processing of ores and revolutionised copper production.[29] A separation method which spread very quickly was the flotation technique. It was used for the first time in 1901 in New South Wales, Australia, in the production of zinc concentrates.[30] This method eventually dominated world ore processing and was, by the early twentieth century, used on multiple low-grade ores, notably copper and gold. The use of electrolysis in metallurgy has its origin from James Elkington, an English plater. He invented a process around 1850, which refined copper using electricity.[31] The first electrolytic refinery was established in 1869 at Pembrey, near Swansea, in the copper smelting plant of Mason & Elkington. In Swansea, large quantities of blister copper were imported from Spain and South America and further refined with the use of electrolysis. Magnetic separation was invented by John Price Wetherill and used in Cornwall and Tasmania on wolframite, but later also used on different types of metals.[32]

In 1899, Daniel C. Jackling, an American mining and metallurgical engineer, and Robert C. Gemmell, an American engineer, wrote a plan on how to mine and mill ore at the rate of 2000 tons per day, which rationalised mining of low-grade disseminated copper ore. The new techniques included the use of mechanical shovels driven by electric power to load mineral, pneumatic drills, rail transport, and blasting to remove large mineral blocks, and by 1910, their idea was realised in

---

[28] R. Chadwick, "New extraction processes for metals" in Singer, C. et al. (eds.). *A History of Technology, volume V The Late Nineteenth Century 1850 to 1900.* Oxford, 1958, p. 82.

[29] "The Copper Industry", *Colonist*, Volume XXVIII, Issue 4163 (12 June 1885), p. 4.

[30] R. Chadwick, "New extraction processes for metals" in Singer, C. et al. (eds.). *A History of Technology, volume V The Late Nineteenth Century 1850 to 1900.* Oxford, 1958, p. 75.

[31] W. H. Dennis, *Metallurgy 1863–1963* (London, 1963), pp. 10 and 28.

[32] R. Chadwick, "New Extraction Processes for Metals", in *A History of Technology, volume V The Late Nineteenth Century 1850 to 1900,* Singer, C. et al. (eds.) (Oxford, 1958), pp. 74 and 85.

open-pit copper mines in Bingham, Utah.[33] The method proved that if the scale was large enough, low-grade disseminated copper ores of only two per cent copper could be extracted with profit.[34] These techniques were key to carrying out large-scale copper mining.

Such methods, equipment, and work systems were used to rationalise mining projects worldwide, but how were new and updated mining and processing devices, furnaces, converters, and methods transferred to and used properly in Norway? Chapter 2 shows that Norwegian mining went through radical technological changes in the late nineteenth and early twentieth centuries, based largely on the adoption of new and more efficient power sources, processing techniques, tools, and machinery. How was technology transferred, and how can we characterise the knowledge on which technology transfer was based? The economist Bengt-Åke Lundvall's categorisation of knowledge into "know-what", "know-why", "know-who", and "know-how" is useful when seeking to detect different aspects of technology transfer processes. Know-what is the factual information about the technology, such as published information about new technology and new updates about the industry in technical magazines and journals. Articles and notes included advertisements about new mining devices, converters, furnaces, and equipment, and were a good starting point for mining companies to acquire an idea of the technology that existed on the market. In Norway, the mining journal *Miner* was published in 1846 and 1847, once a month, and included some descriptions of mining techniques. The *Mining Journal* was published every month from 1913 by the Norwegian Mining Engineer Association. Some of the published articles included detailed descriptions of techniques and functions of specific machines, which served largely as "user recipes". In September 1913, an article about "setz machines" described in detail how these were used in separation processes, how the machines were used in operation, on which minerals to use them, and so on.[35] Moreover, travelogues written by visitors to industrial

[33] S. Villalobos et al. *Historia de la ingenieria en Chile* (Santiago, 1990), p. 174.

[34] J. Temple, "Metal Mining", *A History of Technology, volume VI The twentieth century 1900 to 1950 part 1*, in Williams, T (ed.), (Oxford, 1978), p. 413.

[35] Bergingeniørforening, *Tidsskrift for Bergvæsen* (Oslo, 1913), pp. 2 and 102–104.

exhibitions were sometimes published in magazines. Mining engineer Holm Holmsen participated in the General Art and Industrial Exposition in Stockholm in 1897 and made a report about the mining section in the *Technical Magazine*, including descriptions and drawings of Swedish mines, drilling machines, hydraulic lifts, other lifts, safety facilities, and tools.[36] In this way, information about new technology was diffused. All members of the Norwegian Mining Engineer Association and Norwegian Mining Industry Association had access to the *Mining Journal*, and foreign magazines which also included updated information about new technology.[37]

Yet, reading about new technology in magazines was not enough to fully understand how technology functioned. Nathan Rosenberg explains that "production function as a "set of blueprints" comes off very badly if it is taken to mean a body of techniques which is available independently of the human inputs who utilize it".[38] It was therefore important to acquire the practical skills and capabilities to use the techniques, defined here as know-how. To be able to transfer and adopt machinery and equipment properly it was necessary to observe them in operation. Visits to international exhibitions fulfilled this function to a certain degree as they were important sources of information about up-to-date machinery, equipment, and furnaces. They were used to discover, observe, and assess new technology on the market and there are traces of at least nine engineers going to industrial exhibitions in Norway and abroad between 1881 and 1914 with the purpose of learning about mining techniques. Theodor Wilhelm Holmsen, for instance, went to the Universal Exposition in Paris in 1900 to study "modern mining", especially modern pumps and lifting machinery.[39] Johan Herman Lie Vogt, a well-known geologist and mining engineer, participated in numerous foreign exhibitions, notably the Universal Expositions in Paris in 1889 and 1900, as well as the International Exhibition of Mining and Metallurgy in London in 1890.[40]

---

[36] *Teknisk Ukeblad* (Kristiania, 26 August 1898), p. 465.

[37] Bergingeniørforening, *Tidsskrift for Bergvæsen* (Oslo, 1913), p. 1.

[38] N. Rosenberg, *Perspectives on Technology* (London, New York and Melbourne, 1977), p. 155.

[39] *Studenterne fra 1889* (Trondhjem, 1914), pp. 103–104.

[40] *Studenterne fra 1876* (Kristiania, 1901), pp. 266–275.

Then again, observing models at industrial exhibitions did not alone enable transfer and adoption of technology. To capture the "tacitness" of how to use machinery, equipment, tools, power stations, management systems, and so on, their strengths and weaknesses, and to understand all dimensions of how to select, transfer, adopt, and modify these to suit local conditions, a deeper understanding of their practical use was attained by visiting sites, companies, educational establishments, and research centres where they were already in use. An example is Jacob Pavel Friis' visit to different companies in Sweden in 1876 to learn about the use of coke in copper smelting on behalf of Røros Copper Works. After learning about the use of coke in Åtvidaberg Copper Works he convinced the Board of Directors to adopt this method in Røros. He concluded saying that "[a]fter what I saw and heard about the use of coke in the smelting of matte, it was quite clear that coke might also be used in smelting at Røros".[41] Friis also acquired contacts during this trip. Know-who is the knowledge about who the relevant people are for the solution of problems, and networking and acquiring contacts with experts in Norway, and abroad, were an important part of the innovation processes.[42] Know-why is defined as the scientific and technological principles for the solution of problems, and would require longer theoretical and practical experience, either in Norway or abroad. It is evident that for mining companies to adopt a new machinery, tool, or equipment successfully, they needed to (1) have information about the kind of technology that existed, (2) know who had relevant information about the technique in question, (3) know how it worked in operation, and (4) know why it worked the way it did to understand how it could be changed if further modifications and adaptations were needed. Key to transferring technology was to have comprehensive theoretical and practical knowledge of how it functioned and how it could be adapted to local conditions.

The complexity of technology transfer processes, and how different aspects of knowledge, here identified as know-what, know-who, and know-how, were involved, can be exemplified with the adoption of the

[41] J. P. Friis, *Direktør Jacob Pavel Friis' erindringer* (Oslo, 1944), pp. 173–174.

[42] B. Å. Lundvall, *From the Economics of Knowledge to the Learning Economy* (OECD, 2000).

Manhés process at Røros Copper Works. In the year 1886, the Board of Directors of Røros Copper Works contemplated using the Bessemer method for copper ore processing. Information about the technique—know-what—was collected from "various sources", and the Board considered it to be a good alternative to the technique that was in use.[43] The next step was to find out how the Bessemer method functioned in practice and to find people with relevant know-how. The same year, Anton Sophus Bachke, a member of the Board of Directors, planned a trip to multiple European countries to acquire further know-how about new techniques on the market. The Manhés process, based on Henry Bessemer's iron-processing technique, was used in Swansea by the "great copper king" Sir Hussey Vivian. Bachke contacted Vivian, who invited Bachke and presented the Manhés furnace and its uses. This practical experience seemed to be essential in the selection process. In February 1887, after a long assessment process, the Board decided that the Manhés process would be a good alternative.[44] Yet, multiple adaptations were necessary to put the method in use and through contacts—know-who—a foreman (contremaître) was sent from France to Røros in October 1887 to assist with the installation of the new furnaces and equipment.[45] Understanding how long the matte should smelt in the water-jacket ovens, pouring the right amount of matte into the converters, making the air blow correctly into the converters, carrying out the process efficiently, and knowing what the concentrated product should look like when it was finished were complex tasks, and practical assistance from an expert in the field was probably very useful. After a number of tests, failed attempts, and further modifications, the process was in use from October 1888 with good results.[46]

The chemist Olav Steen's work abroad demonstrates how drawing on long practical work experiences in other countries was useful when

[43] Røros kobberverk, *årlig rapport 1886*, p. 9.

[44] Røros kobberverk, *årlig rapport 1887*, p. 8.

[45] H. Dahle, *Røros Kobberværk 1644–1894* (Trondheim, 1894), p. 447.

[46] Bergingeniørforening, *Tidsskrift for bergvæsen*, 1916, p. 43.

selecting new smelting techniques. Steen began a career in mining after graduation in 1895. During the first years he worked at multiple iron and steel companies in Germany and Italy. One of his working tasks was to travel around Europe and study smelting techniques for iron and steel. In 1917, he returned to Norway and began working as operation manager at Stavanger Electrical Steel Works. Immediately after his return, it was decided to modernise the company. In this regard, Steen argued that the furnace that was used at the company, a Roechling-Rodenhauser induction furnace of four steps, was not safe. He found that it had low durability, long repair time, and was expensive to operate. He recommended electric arc smelting instead, which he had studied and worked with on the Continent. He suggested an eight–ten-ton Heroult furnace, and a gradual reduction of the induction smelting furnace, which would eliminate coal and lead to important cost reductions. This furnace was similar to the one he had installed at an iron works in northern Italy a couple of years earlier. The Directory must have been convinced, because the furnace he recommended was ordered from England. After a couple of years, it was installed and used successfully in operation.[47] It seems like Steen's long practice with iron and steel smelting techniques abroad contributed to his understanding of how they functioned and about underlying principles on how they operated, and this experience, in turn, turned out to be of great advantage to Stavanger Electrical Steel Works.

Summary: Use of Complex Knowledge

Mining, and transfer, modification, installation, and adoption of technology, was based on complex knowledge. Natural sciences—notably geology, mineralogy, chemistry, metallurgy, mechanics, physics, and mathematics—were key to measuring, designing, and making of mines

[47] *Studentene fra 1890*, pp. 217–218; Oslo tekniske skole, *Skrift ved 50 årsjubileet*, pp. 92–102.

as well as analysing and processing ore. Mapping, surveying, planning, drawing, and constructing underground mines involved complex calculations, measurements, and structures without which an in-depth comprehension of theoretical knowledge within these scientific disciplines would have been extremely difficult. Constructions and structures beneath the earth were particularly challenging, since the mines could easily collapse. Intricate mathematical calculations were required to make sustainable, deep, and large mines. At the same time, mining depended largely on specific knowledge of the local geological ground and was heavily based on physical and practical tasks. The practicality of mining suggests that the activities relied to a large degree on knowledge which might be hard to explain, that is tacit knowledge. The tacit dimension seemed to be present in all mining activities and much learning thus happened through observing, doing, and hands-on practice.

Radical technological changes occurred in world mining in the late nineteenth and twentieth centuries to cope with new market conditions and geological settings, and to ensure rational and profitable mining projects. The exhaustion of high-grade ores and the making of deeper and bigger mines in more remote areas increased the need for detailed geological maps, ore analyses, and economic planning. Mechanic power replaced animals and man force and called for in-depth knowledge in mechanical and construction engineering. The use of electricity in mining operations from the late nineteenth century and new electro-chemical and metallurgical methods required new specialised knowledge in chemistry, metallurgy, and electro-engineering. Mineral and metal processing became increasingly complex and specialised for at least two reasons. First, the extraction of lower-grade ores often required different processing techniques than those for high-grade ore. Second, the extraction of new mineral and metal products, such as aluminium, molybdenum, and steel, demanded new knowledge-based experiments. Furthermore, companies grew in size and focused increasingly on adopting more efficient management and coordination systems. Knowledge in business administration, accounting, and economics became crucial.

In Norway—a catching-up economy of the Industrial Revolution—the adoption of new technology was to a large degree based on transfers from abroad. Companies were kept informed about new and more efficient techniques, systems, equipment, furnaces, converters, and so on in different ways, but in order to select, transfer, install, and use new equipment, furnaces, and converters, and to adapt them to local settings successfully, in-depth and comprehensive knowledge of how they functioned in operation was key. This required more than reading about updates in magazines and technical journals and often involved travelling to plants where new technology was in use. The knowledge categorisation made by Bengt-Åke Lundvall—know-what, know-who, know-how, and know-why—is useful here. Know-what is defined as the factual information about the technology; know-who is the knowledge about who the relevant people are for the solution of problems; know-how is the ability, practical skills, and capabilities to do something; and know-why is the scientific and technological principles for the solution of problems.[48] Together, these elements seemed to capture the broad set of knowledge that was involved in technology transfer processes. It should be mentioned here that it was not only in mining that innovation largely involved knowledge transfer from abroad. Kristine Bruland shows that the textile industry in Norway developed largely based on the ability among Norwegians to adapt, use, imitate, and copy foreign machines. Similarly, the development of the workshop industry in Norway was a result of "training and education, access to information on foreign technical advances and the ability to use information".[49] The dependence on foreign technology was, thus, a wide phenomenon.

In sum, mining was based on, and used, a broad spectre of knowledges. Table 3.2 provides a simple overview of knowledge areas that were used in finding, analysing, mining, and extracting ore into pure metal and mineral products and alloys.

---

[48] B.-Å Lundvall, Bengt-Åke. "Post Script: Innovation System Research – Where It Came From and Where It Might Go". In *National Systems of Innovation*, B.-Å. Lundvall (ed) (UK, 2010), 330.

[49] K. Bruland, "Norsk mekanisk verkstedindustri og teknologioverføring 1840–1900", in *Teknologi i virksomhet: verkstedindustri i Norge etter 1840*, E. Lange (ed.) (Oslo, 1989), pp. 55–60 and 73.

Table 3.2 Simple overview of knowledge used in technologically advanced mining

Steps in the mining process	Scientific knowledge areas	New knowledge areas from the late nineteenth century	Tacit knowledge	Transfer of new machinery, equipment, converters, furnaces, etc.
1. Geological surveys/ prospecting	Geology Mineralogy Chemistry Economics			
2. Removal of ores	Physics Mathematics Construction engineering Mechanics	Electro-engineering Economics Administration	High degree	Continuous acquisition of Know-how Know-what Know-who Know-why
3. Processing of ores	Geology Mechanics Chemistry Mineralogy Metallurgy (electro-engineering, electro-chemistry, etc.)			

Sources: H. Carstens, *...Bygger i Berge: en beretning om norsk bergverksdrift* (Trondheim, 2000); F. Sæland, *Bergingeniør Emil Knudsens erindringer Inntrykk fra et bergmannsliv 1856–1897* (Kongsberg, 2005); B.-Å Lundvall, "Post Script: Innovation System Research – Where It Came From and Where It Might Go". In *National Systems of Innovation*, B.-Å. Lundvall (ed) (UK, 2010); J. P. Friis, *Direktør Jacob Pavel Friis' erindringer* (Oslo, 1944); H. Dahle, *Røros Kobberværk 1644–1894* (Trondheim, 1894); Bergingeniørforening, *Tidsskrift for Bergvæsen* (Oslo, 1913–1940); R. Chadwick, "New Extraction Processes for Metals", in *A History of Technology, volume V The Late Nineteenth Century 1850 to 1900,* Singer, C. et al. (eds.) (Oxford, 1958)

Bibliography

Secondary Sources (Literature)

Bergingeniørforening. (1913). *Tidsskrift for kemi og bergvæsen*. Oslo: Den norske ingeniørforeigning og Den polytekniske forening.

Bergingeniørforening. (1916). *Tidsskrift for kemi og bergvæsen*. Oslo: Den norske ingeniørforeigning og Den polytekniske forening.

Bergingeniørforening. (1925). *Tidsskrift for kemi og bergvæsen*. Oslo: Den norske ingeniørforeigning og Den polytekniske forening.

Bergingeniørforening. (1926). *Tidsskrift for kemi og bergvæsen*. Oslo: Den norske ingeniørforeigning og Den polytekniske forening.

Bergingeniørforening. (1937). *Tidsskrift for kemi og bergvæsen*. Oslo: Den norske ingeniørforeigning og Den polytekniske forening.

Boletín de la Sociedad Nacional de Minería. (1925). Santiago.

Boletín de la Sociedad Nacional de Minería. (1928). Santiago.

Børresen, A. K., & Wale, A. (red.). (2005). *Vitenskap og teknologi for samfunnet? Bergfagene som kunnskapsfelt*. Trondheim: Tapir akademisk forlag.

Carstens, H. (2000). *...Bygger i Berge: en beretning om norsk bergverksdrift*. Trondheim: Norsk bergindustriforening Den norske bergingeniørforening Tapir.

Chandler, A. (1990). *Scale and Scope: The Dynamics of Industrial Capitalism*. Cambridge, MA: Harvard University Press.

Dahle, H.. (1894). *Røros Kobberværk 1644–1894*. Trondheim.

Dennis, W. H. (1963). *Metallurgy 1863–1963*. London: Routledge.

Foray, D., & Lundvall, B.-Å. (1996). *From the Economics of Knowledge to the Learning Economy*. Paris: OECD.

Friis, J. P. (1944). *Direktør Jacob Pavel Friis' erindringer*. Oslo: Cammermeyers Boghandel.

Hirsch-Kreinsen, H., et al. (2003). *Low-Tech Industries and the Knowledge Economy: State of the Art and Research Challenges, PILOT Policy and Innovation in Low-Tech*. Oslo: STEP – Centre for Innovation Research.

Hodne, F., & Grytten, O. H. (2000). *Norsk økonomi i det nittende århundre*. Bergen: Fagbokforlaget.

Hunstadbråten, K. (1997). *Blaafarveværket*. Drammen: Brakar A/S.

Imsen, S., & Winge, H. (1999). *Norsk historisk leksikon* (2nd ed.). Oslo: Cappelen Akademisk Forlag.

Lange, E. (Ed.). (1989). *Teknologi i virksomhet: verkstedindustri i Norge etter 1840*. Oslo: Ad notam forlag.

Lov, ang. Røraas Kobberværk: [af 12. Sept. 1818, med tillæslove]: Trondhjem.

Lundvall, B.-Å. (1992). *National Systems of Innovation*. London: Pinter Publishers.

Norges offisielle statistikk. (1940). *Norges Bergverksdrift*. Oslo: Statistisk sentralbyrå.

Orkla Grube-Aktiebolag. (1954). *Løkken Verk En Norsk Grube Gjennom 300 År*. Trondhjem: Orkla Grube-Aktiebolag.

Oslo tekniske skole. (1944). *Skrift ved 50 årsjubileet for ingeniørene fra KTS 1894*. Oslo.
Røros kobberverk. (1886). *årlig rapport 1886*. Trondheim: Røros kobberverk.
Røros kobberverk. (1887). *årlig rapport 1887*. Trondheim: Røros kobberverk.
Rosenberg, N. (1977). *Perspectives on Technology*. London/New York/Melbourne: Cambridge University Press.
Singer, C., et al. (Eds.). (1954–1975). *A History of Technology*, 5 volumes. Oxford: The Claredon Press.
Sogner, K. (2003). *Elkem gjennom 100 år Skaperkraft 1904–2004*. Oslo: Messel Forlag.
Studentene fra 1890. (1942). Oslo: Grøndahl & Søn Boktrykkeri.
Studenterne fra 1876. (1901). Kristiania.
Studenterne fra 1889. (1914). Trondhjem.
Sæland, F. (2005). *Bergingeniør Emil Knudsens erindringer Inntrykk fra et bergmannsliv 1856–1897*. Kongsberg: Norsk bergverksmuseum.
Teknisk Ukeblad. (1895). Kristiania: Den norske ingeniør- og arkitektforening og Den polytekniske forening.
The Colonist. (1885, June 12). Nelson. *XXVIII*(4163).
Villalobos, S., et al. (1990). *Historia de la ingenieria en Chile*. Santiago: Editorial Universitaria.

Part III

A Historical Empirical Analysis of Knowledge Organisations

The previous part established that the productive and changing mining sector in Norway was based on complex knowledge, which was identified as knowledge taken from natural sciences, engineering, economics, and management, as well as tacit and practical knowledge. The categorisation of knowledge into know-what, know-how, know-who, and know-why seems to largely cover the aspects of knowledge involved in technology transfer processes. But how was such useful knowledge for mining attained and ensured? In previous chapters—in given examples of transfer and use of technology and knowledge—multiple knowledge organisations have been mentioned, notably mining companies, technical societies, laboratories, universities, and research centres. For example, information, or know-what, was provided in magazines published by societies and combined with hands-on practical knowledge of techniques acquired at mining companies. In this part of the book I explore how knowledge organisations—in Norway, but often connected to foreign organisations—developed and accumulated knowledge within natural sciences, and tacit, practical knowledge and know-what, know-how, know-who, and know-why.

In general terms, a whole set of organisations were potentially involved in the start-up and advance of mining projects. A whole collection of domestic and foreign companies, universities, laboratories, industrial exhibitions, professors, consultants, and engineers interacted, and collaborated, with each other in processes of developing technological capabilities and accumulating, transferring, and using technology. I look here to (1) universities and mining schools providing formal mining education; (2) mining companies, both domestic and foreign businesses; (3) research centres, notably the Geological Survey of Norway; (4) mechanical workshops (the capital goods industry) providing mining equipment, machinery, and technical services to a number of industries, including mining. Did these knowledge organisations contribute to the technological transformation that the mining sector went through? If so, how? What were their functions? Were there any limitations to the knowledge they supplied?

4

The University, the Norwegian Institute of Technology (NIT), Technical Schools, and the Mining School

There was an increased focus on science in the eighteenth century, which was accompanied by the establishment of formal mining education. Mining education was one of the first technical training programmes aimed to prepare engineers and technicians for a specific industry. The first mining school in Europe was created in Freiberg, Germany, in 1702, but schools of mining and metallurgy were also established in Austria, Russia, France, Italy, Norway, and Sweden in the early decades of the eighteenth century. The first School of Mines in the United States was the School of Mines of New York, which was established in 1864. The first mining school in Latin America was the Mining School of Copiapó in Chile, founded in 1857, and other mining schools were opened in Colombia, Bolivia, and Brazil in the late nineteenth and early twentieth centuries. The Mining and Engineering College was established in Wuchang in China in 1892 and the Imperial College of Engineering in Japan, founded in 1875, offered courses in geology, mining, and metallurgy.[1]

Heinrich Schlanbusch, a German mining engineer, suggested in 1687 to create a public mining seminar in Norway, but it was not until 70 years later, in 1757, after another recommendation by one of the managers at Kongsberg

[1] F. Habashi, *Schools of Mines* (Quebec, 2003), pp. 115–422, 450, 516, and 521.

K. Ranestad, *Knowledge-Based Growth in Natural Resource Intensive Economies*, Palgrave Studies in Economic History, https://doi.org/10.1007/978-3-319-96412-6_4

Silver Works, that the Mining Seminar in Kongsberg was established. It was funded by the state.[2] The state continued to finance the formal mining engineering education, and in 1814, the programme was transferred from Kongsberg to the newly opened Royal Frederick University in Christiania.[3] In 1914 it was moved once again to the newly opened Norwegian Institute of Technology (NIT) in Trondheim, which was largely based on the German technical educational system, and which provided engineering programmes on a tertiary level.[4] A public intermediate mining school—Kongsberg Silver Works Elementary Mining School—was established in Kongsberg in 1867, which meant that, from then on, formal mining education was provided on a higher as well as on an intermediate level in Norway.[5]

Moreover, three intermediate technical schools in Bergen, Trondheim, and Oslo were established in the 1870s and offered a variety of technical study programmes, such as mechanics, chemistry, electro-engineering, and carpentry, which might be considered useful for the increasingly technologically complex mining sector. Other technical schools and evening schools were founded in various towns and near industrial areas, such as Bergen Drawing School of 1772, Horten Technical School of 1855, Skienfjorden Technical School of 1887, and the School of Agriculture of 1854 (Higher School of Agriculture from 1897).[6] Technical evening schools were founded in cities such as Stavanger and Kristiansand in the late 1870s.[7] The NIT, in addition to mining engineering, offered engineering programmes in architecture, electrical engineering, mechanical engineering, construction engineering, and shipbuilding.[8] The questions addressed here are the following: did the educational establishments develop useful and relevant knowledge for mining? If so, what kind of knowledge, and how was it used?

---

[2] B. I. Berg, *Gruveteknikk ved Kongsberg Sølvverk 1623–1914* (Trondheim, 1998), p. 167.

[3] Old name for Oslo.

[4] T. J. Hanisch and E. Lange, *Vitenskap for industrien NTH – En høyskole i utvikling gjennom 75 år* (Oslo, 1985), p. 23.

[5] Statens bergskole, *Bergskolen 100 år Jubileumsberetning 1866–1966* (Trondheim, 1966), p. 9.

[6] T. Bergh et al. *Norge fra u-land til i-land: vekst og utviklingslinjer 1830–1980*, (Oslo, 1983), p. 52.

[7] F. Hodne, *Norges økonomiske historie 1815–1970* (Oslo, 1981), p. 245.

[8] T. J. Hanisch and E. Lange, *Vitenskap for industrien NTH – En høyskole i utvikling gjennom 75 år* (Oslo, 1985), p. 23.

Mining Engineering and Technician Study Programmes

In the eighteenth century, the main purpose of the Kongsberg Mining Seminar was to educate engineers for mining companies. After the mining engineering programme was moved to the University, the aim was twofold: first, to provide the mining industry with capable managers for mining companies and, second, to provide scientists to the University and research centres. The University took a scientific approach and the instruction included more theoretical teaching than the one at the Mining Seminar. Nevertheless, the practical training was not completely taken out, and was offered to the students at the state-owned Kongsberg Silver Works.[9] What courses did the mining engineering and technician study programmes entail? Were they useful for mining? Were other study programmes useful for mining? If so, how?

The mining engineering programme was set to four years in 1814. It was later extended to five years. The basis of the mining engineering programme was natural science courses, notably mathematics, mechanics, geology, and mineralogy, which lay the foundation for key knowledge of how to design, plan, and construct mines.[10] In addition to in-depth teaching of natural sciences, the programme included instruction of "mining" itself. This involved courses which were more directed towards specific tasks in mining. Some of these courses included the teaching of how to maintain and manage mines and how to use and repair mining machinery. These courses were "mining construction", "mine factory", and "machine drawing".[11] Another mining course was metallurgy. In this course the students were introduced to metal characteristics and extraction methods.[12] This course was probably highly relevant for the mining engineering students considering that more metallurgical specialisations developed during the nineteenth century, more processing techniques were introduced on the market, and more metals and minerals were being produced.

[9] Bergingeniørforening, *Tidsskrift for kemi og bergvæsen* (Oslo, 1932), p. 205.

[10] See the formal mining engineering programme from the University in K. Ranestad, "The mining sectors in Chile and Norway", thesis, Geneva, 2015, pp. 444–449.

[11] UIO, *Universitets- og skoleannaler* (Kristiania, 1834–1835), p. 215.

[12] A. K. Børresen, *Bergtatt: Johan H. L. Vogt* (Trondheim, 2011), pp. 67–68.

Additional courses which focused on mining machines and equipment were adopted eventually. In 1871, "study of machines" became a course and object of examination.[13] This change reflects the mechanisation process in mining which began around the mid-nineteenth century and the need to know how to use and repair steam engines, lifts, mechanical devices, and so on. In response to the adoption of electric power by the turn of the century, electro-engineering was added as a separate course in 1909.[14] Inorganic chemistry was added as a specialised course in chemistry in 1910, which included lectures about chemical elements, physical chemistry, electro-chemistry, and mechanical technology.[15] Electro-metallurgy was included in the programme at the University in 1913 and responded to the large-scale mineral and metal productions based on ore extraction that were initiated around the turn of the century.[16]

The programme seemed to adapt to and follow general technological changes that were made. Modifications and adoption of new courses occurred normally *after* new equipment, methods, or techniques had been adopted and disseminated throughout the sector. For example, the new course in mining machinery was created after mechanised equipment had been adopted by mining companies and the course in electro-engineering was introduced after the shift to electric power. All in all, the mining engineering programme covered a broad spectre of knowledge areas and seemed to roughly match the natural sciences knowledge areas, and new knowledge areas, on which mining was based. The variety of courses covered by the mining engineering programme, and the adaptations that were made with the introduction of new and specialised courses, suggest that the students were given a solid scientific and theoretical knowledge foundation, which was both relevant and useful for mining in Norway (see Table 4.1).

The increased diversity and complexity, and the specialisations which characterised mining towards the late nineteenth century were important arguments for an extensive mining engineering programme. In the late

---

[13] UIO, *Det Kongelige Norske Frederiks Universitets Aarsberetning* (Kristiania, 1872), p. 149.

[14] Bergingeniørforening, *Tidsskrift for kemi og bergvæsen* (Oslo, 1932), p. 134.

[15] NTH, *Program for studieåret 1912–1913* (Trondhjem, 1913), pp. 25 and 41–43.

[16] Bergingeniørforening, *Tidsskrift for kemi og bergvæsen* (Oslo, 1932), p. 134.

Table 4.1 Simple overview of knowledge areas and the mining engineering study programme

Scientific knowledge areas	New knowledge areas from the late nineteenth century	Mining instruction	
		Continuous courses from 1814:	**New courses (first adopted):**
Mathematics	Electro-engineering	Mathematics	Study of machines: 1871
Physics/mechanics	Economics	Mechanics	Electro-engineering: 1909
Geology	Administration	Geology	House construction: 1911
Mineralogy		Mineralogy	Social economics and law: 1911
Chemistry		Mining construction	
Metallurgy		Mine factory	
Construction engineering		Metallurgy (ore treatment and analyses)	
		Machine drawing	

Sources: NTH, *Program for studieåret 1912–1913* (Trondhjem, 1913); NTH, *Program for studieåret 1919–1920* (Trondhjem, 1920), p. 23; NTH, *Beretning om virksomheten 1910–1920* (Trondhjem, 1920); UIO, *Det Kongelige Norske Frederiks Universitets Aarsberetning* (Kristiania, 1872); UIO; *Universitets- og skoleannaler* (Kristiania, 1834–1910)

eighteenth century—during the time when the Mining Seminar in Kongsberg provided the formal mining education in Norway—mining seemed to be less based on scientific principles and more dominated by traditions and craft skills.[17] After the turn of the century, formal education was viewed as essential to the advancement of the sector and even became mandatory for certain managing positions. In 1818, Røros Copper Works decided by law that the director of the company had to have formal training in mining: "As Managing Director at Røros Copper Works must in the future no one be selected, who have not fully studied mining, and passed exams in his theoretical as well as practical skills in mining engineering." The other middle managers in the company, including mine managers, should have knowledge of mining construction, ore surveying and mineralogy, and ore dressing and metallurgy.[18] In 1833,

[17] B. I. Berg, *Gruveteknikk ved Kongsberg Sølvverk 1623–1914* (Trondheim, 1998), 314.

[18] *Lov, ang. Røraas Kobberværk* (Trondhjem, 12 September 1818), p. 17.

Jacob Aall, Member of Parliament and owner of Næs Iron Works, wrote that the metal works were the country's most important factories, and thus proper management demanded "numerous kinds of knowledge and a long preparation".[19] In 1910, the NIT wrote that "[a] large modern mining enterprise is a technically very complicated business and requires among engineers and managers a significant insight in numerous and extensive areas".[20] This growing conviction that mining education was important should be understood in relation to the ongoing technological changes that were occurring in terms of applying new and more efficient energy sources, new techniques for finding, removing, and processing ore, new management systems, and adopting large-scale production.

The belief in mining engineers as key for the development of mining businesses is observed in many countries from the late nineteenth century. It is found that "a broad general academic education with some basic science was essential" for iron mining in Germany and Britain. Especially managers in these industries needed a "sound basic education which included science".[21] In Australia and Chile too—countries with a long mining tradition and increasing production of metals and minerals in the early twentieth century—mining engineers were understood as crucial.[22] The Director of the Mining and Saltpetre School in Antofagasta, Chile, wrote in 1926 that without theoretically prepared mining technicians and engineers, mining operations would fail due to "lack of a conscious, rational and scientific direction".[23] There was a strong focus on mining education in the United States, which had more than 20 schools which granted degrees in mining in the period between 1860 and 1890, and in 1893 the country "had more mining students than any country in Europe, except Germany".[24] More specifically, Samuel Christy reported

[19] Bergingeniørforening, *Tidsskrift for kemi og bergvæsen* (1932), p. 133.

[20] Norges Tekniske Høiskole, *Beretning om virksomheten 1910–1920* (Trondhjem, 1920), 56.

[21] P. Musgrave, *Technical Change the Labour Force and Education* (Oxford, New York, 1967), pp. 27 and 74.

[22] S. Villalobos et al. *Historia de la ingenieria en Chile* (Santiago, 1990); S. Macchiavello Varas, *El problema de la industria del cobre en Chile y sus proyecciones económicas y sociales* (Santiago, 2010).

[23] *Boletín de la Sociedad Nacional de Minería* (Santiago, 1926), pp. 644–646.

[24] P. David, G. and Wright, "Increasing Returns and the Genesis of American Resource Abundance", *Oxford Journals*, vol. 6, issue 2 (1997), p. 231.

at the International Engineering Congress in August 1893 that mining engineers "may be insignificant in numbers, but in respect of the value produced as a result of their labor, they are the most important element in the entire population".[25]

In 1914, the mining engineering programme in Norway was transferred to the newly opened NIT. The study programmes offered there were as long as the ones at the University with a focus on scientific theory. Instruction was provided on a tertiary level.[26]A Committee was established to discuss the future of the mining engineering programme and it quickly concluded that the programme be moved to the NIT since the technical courses, such as machine study and construction studies, electro-engineering, and surveying, were more developed at the NIT than at the University. Teaching facilities were new, and the laboratories were well equipped.[27]

The NIT maintained courses in natural sciences as the basis for the mining engineering programme, but at the same time an increased focus was directed on courses in "machinery" and "construction". By 1914, most mining companies used technical equipment and complex constructions, such as drilling, draining, ventilation, and transport systems, ore dressing, and extraction equipment, which explains the increased emphasis in this area.[28] Courses in bookkeeping and economics were also added to the programme. The course on "social economy and law" included topics about the relation between capital, work, and nature, increase in value, capital interest, and wages. A separate course was offered in bookkeeping from 1913 and finance and statistics were introduced as separate courses in the 1920s.[29] The new focus on management, accounting, and economic planning is probably explained by the

[25] S. B. Christy, "The Growth of American Mining Schools and Their Relation to the Mining Industry", Chicago Meeting, part of the International Engineering Congress (California, August 1993), pp. 453–454.

[26] E. Benum, "Dannelse", "Tillid" og "Autoritet", in *Makt og Motiv Et festskrift til Jens Arup Seip*, J. A. Seip and O. Dahl (Oslo, 1975), pp. 130–131.

[27] Norges Tekniske Høiskole, *Beretning om virksomheten 1910–1920* (Trondhjem, 1920), p. 56.

[28] NTH, *Beretning om virksomheten 1910–1920* (Trondhjem, 1920), p. 56.

[29] NTH, *Program for studieåret 1913–1914* (Trondhjem, 1914), pp. 34–35; NTH, *Program for studieåret 1919–1920* (Trondhjem, 1920), p. 23; NTH, *Beretning om virksomheten 1910–1920* (Trondhjem, 1920), pp. 54–55.

increased size of companies from the late nineteenth century (see Table 4.2 of the study programme at NIT from 1920).

Know-how in management, accounting, and economic planning was also called for other in other countries with industries based on metal and mineral mining and extraction. In Britain, it is found that larger units and workforce from the turn of the century required increased skills in business and management.[30] In the United States, mining became increasingly intricate due to new mining methods, new devices, more complex interaction among parts of the system, and a growing workforce, which is shown to have led to the use of new technical skills, as well as new managerial and economic skills.[31] In Chile—a country with large foreign corporations investing in mining from the late nineteenth century—the University of Chile added new courses in "administration" and "economics" to the mining engineering programme in 1889 and 1908.[32]

In Norway, in response to the diversification, and the branching out of large-scale production of iron, steel, aluminium, and so on, and the many new ore-crushing and ore-processing techniques on the market, the mining programme was divided in two in 1931, with one sub-programme specialising in "mining" and another in "metallurgy". The two study programmes included many of the same courses, but the first one included more teaching of surveys and ore dressing and the second one included more courses in chemistry and metallurgy.[33] This division facilitated career specialisations for the mining engineer students, but it happened decades after these changes were a fact; thus, rather late.

The mining technician programme at the Kongsberg Silver Works Elementary Mining School, which opened in 1867, also adapted to technological changes in mining, and seemed highly relevant and useful for the sector. The intermediate mining technician instruction was organised in a similar manner as the mining engineering programmes and

---

[30] P. Musgrave, *Technical Change the Labour Force and Education* (Oxford, New York, 1967), pp. 134–135.

[31] K. H. Ochs, 'The rise of American mining engineers', *Technology and Culture*. Vol. 33, No. 2 (1992), pp. 282–283.

[32] Universidad de Chile, *Anales de la Universidad de Chile* (Santiago, 1890), pp. 118–119; Universidad de Chile, *Anales de la Universidad de Chile* (Santiago, 1908), pp. 360–361.

[33] NTH, *Program for studieåret 1931–1932* (Trondhjem, 1932), pp. 61–62.

Table 4.2 The mining engineering programme at the NIT in 1920

	Weekly hours							
	1st semester		2nd semester		3rd semester		4th semester	
Courses	Lectures	Exercises	Lectures	Exercises	Lectures	Exercises	Lectures	Exercises
Mathematics I	4	3	4	2				
Descriptive geometry	3	3						
Mechanics I	2	3	2	3				
Mechanics II					4	4		
Physics	4		4					
Exercises in the physical laboratory						3		
Experimental chemistry	2		2					
Inorganic chemistry	4		4					
Mineralogy and geology	3	1	3	4				
Mineralogy and crystallography	2[a]	2			2[a]	2	2	2
Basic features of machine building	2	3	2	3				
"Machine parts"					4	6	3	6
Mechanical technology A					4			2
Surveying[b]				1	3	3	3	3
Qualitative analysis	2	[c]						
Quantitative analysis			1	[c]				
Technical writing		[d]						

	5th semester		6th semester		7th semester		8th semester	
Courses	Lectures	Exercises	Lectures	Exercises	Lectures	Exercises	Lectures	Exercises
Electrical engineering	2		2					
"Descriptive machine teaching"	3	2	3	2	3	2	3	2
Mineral and rock microscopy	3	1			3	1		1
House building I	4			6				

(continued)

Table 4.2 (continued)

Courses	5th semester		6th semester		7th semester		8th semester	
	Lectures	Exercises	Lectures	Exercises	Lectures	Exercises	Lectures	Exercises
Water building	2		2					
"Mining"	3	1	2	1	3	1	2	1
Mineral processing[e]	2	2			2	2		
Mining law[f]			2				2	
Mining measurement and magnetometer			1	1				
Metallurgy	4		3		4		3	
Chemical laboratories and metallurgical exercises		[g]		[g]		[g]		
Social economy and law	2		4		2			
Bookkeeping	1	3						
"Samaritan course"						[h]		

Notes

[a]Joint courses

[b]Hereto 5 weeks practical exercises in the field during holidays

[c]Laboratory exercises by appointment

[d]By appointment

[e]Every other year 1919, 1921, etc.

[f]Every other year 1919, 1921, etc.

[g]By appointment

[h]Around 10 hours per semester

Admission to the 3rd year requires at least a 1-month stay at a mine, and for admission to the final exam a 4-month stay at a mine is required (herein the first month included). Of these 4 months, 2 months must include participation in practical work and at least a 1-month stay at a mine, at least 3 weeks at a crushing plant, and at least 3 weeks at a smelting plant

Source: Den Tekniske Høiskole i Trondhjem. *Program for Studieåret 1919–1920* (Trondhjem, 1920), pp. 46–47

included roughly the same natural science courses and mining courses in "accounting", "machine drawing", "ore surveying", and so on. The difference was that the programme excluded in-depth teaching of scientific theories and was much shorter and much more focused on practical learning (two years; two and a half years from 1870). The instruction focused on short introductions to different knowledge areas and hands-on practice and exercises. In addition to the classroom instruction, the students had to finish at least one year's practice at a mine or metallurgical plant before being admitted to the study programme and they worked at Kongsberg Silver Works during working hours.[34] The theoretical teaching increased to 1560 hours in 1899, and new courses in mining construction, chemistry, and metallurgy were adopted, but practical exercises and daily work at Kongsberg were still an obligatory part of the programme.[35] The objective was primarily to instruct workers from Kongsberg Silver Works, but the school also accepted workers from other mining companies.[36]

In the early twentieth century, formal training became mandatory for mine managers, which were common working positions for mining technician graduates.[37] Mining technicians were normally not meant to administrate and manage companies, but rather to assist engineers in their work. In 1921, the Mining School was closed temporarily; however, the industry expressed its wish to reopen the school again. In 1930, a debate about the future of the Mining School emerged, and the strong argument was that the country still needed mining technicians and foremen with theoretical training. In 1936, the School reopened, and new courses were added, notably "electrical engineering", "ore dressing", "construction and civil engineering", "mining" and "ore surveying", and "the study of company and work".[38] The addition of these courses indi-

---

[34] Statens bergskole, *Bergskolen 100 år Jubileumsberetning 1866–1966* (Trondheim, 1966), pp. 10–11 and 14.

[35] For the study plan, see Ranestad, 'The mining sectors in Chile and Norway', pp. 264–267.

[36] Statens bergskole, *Bergskolen 100 år Jubileumsberetning 1866–1966* (Trondheim, 1966), p. 8.

[37] Bergingeniørforening, *Tidsskrift for kemi og bergvæsen* (Oslo, 1935), p. 154.

[38] Statens bergskole, *Bergskolen 100 år Jubileumsberetning 1866–1966* (Trondheim, 1966), pp. 8–10, 20, and 42.

cates that the School also modified the programme according to the technology that was used in mining. It seems clear that technological changes influenced and shaped the formal mining instruction.

Mining Engineer and Technician Graduates

A total of 341 Norwegian mining engineers graduated between 1787 and 1940. Amongst them 118 (34.6 per cent) studied in another country, 83 studied partly in Norway and partly in another country, while 35 went abroad to study right after high school. Figure 4.1 shows all graduates from 1787 and 1940 and suggests that most of the time there were at least a couple of Norwegian mining students studying abroad. Most of

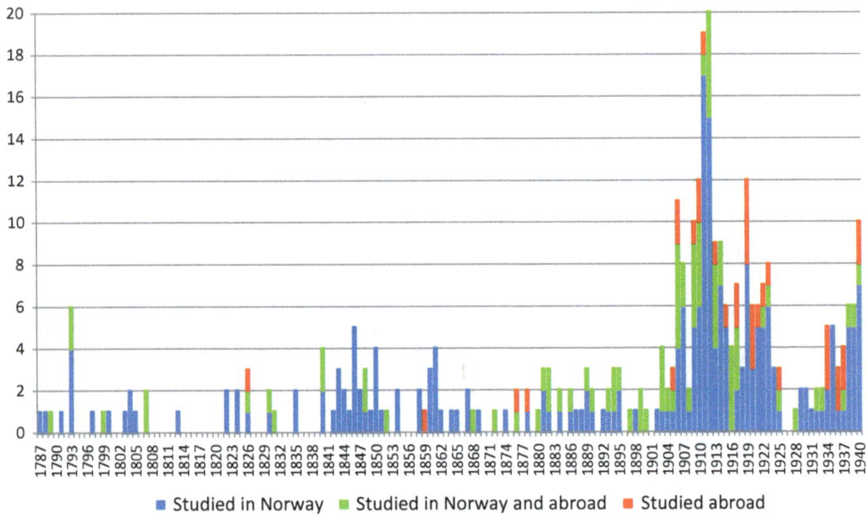

Fig. 4.1 Norwegian mining engineers *Graduated at Norwegian and foreign education institutions* Year of graduation. Sources: *Studenterne* (Oslo, 1855–1940); L. Eskedal, *BTS-matrikkelen Ingeniører uteksaminert ved Bergen Tekniske Skole 1875–1975* (Bergen 1975); Gløersen, *Biografiske Oplysninger*; B. Bassøe, *Ingeniørmatrikkelen Norske Sivilingeniører 1901–1955 med tillegg* (Oslo, 1961); G. Brochmann (red.) *Vi fra NTH de første 10 kull: 1910–1919* (Stavanger, 1934); O. Amundsen, *Vi fra NTH de neste 10 kull: 1920–1929* (Oslo, 1950)

the students studying abroad went with University scholarships, of which there were many (see next chapter).

There were 87 mining engineers who studied in Germany, which was the most popular country to study in apart from Norway. The most popular mining school was Freiberg Mining Academy, which was visited by 48 Norwegian mining engineer students. Other popular destinations were Sweden, Switzerland, France, England, and the United States. Studying abroad—which was encouraged by the University and the NIT—had two main functions. First of all, it meant that students could take courses and learn about subjects which did not exist in Norway. Second, they worked in international scientific environments, sometimes with distinguished professors, which meant that they became acquainted with different scientific traditions and acquired valuable contacts. Emilius Knutsen Looft, for instance, worked as an assistant for a professor in Wiesbaden before he became a mining engineer in 1889.[39] Some of the engineers studied at two or three different universities or schools, and in more than one country. Johan Herman Lie Vogt, for example, visited multiple countries and studied many places. He went to Dresden after high school and studied two semesters at Dresdner Polytechnikum. In 1877, he went back to Christiania and finished his mining studies at the *Royal Frederick University* in 1880. In 1882, he went to Stockholm to study "metallurgy of iron" and "quantitative-chemical analysis" at the Mining School there. At the same time, he worked at the Stockholm University College. In 1884, after multiple excursions in Sweden and Norway, he went to Freiberg in Sachsen where he studied a couple of months at Freiberg Mining Academy. From Freiberg, he went on excursions around Erzgebirge before studying a month at the Mining School in Clausthal and a month at Leipzig University. He travelled to Norway before returning to Stockholm University College and Germany. After short stays in Stockholm and Germany he studied for one and a half months at the Collège de France in Paris.[40] The mining engineers followed a wider Norwegian trend. Before the NIT was established in 1910, the only

[39] *Studentene fra 1884* (Kristiania, 1909), pp. 209–210.

[40] *Studentene fra 1876* (Kristiania, 1901), pp. 66–75.

option for Norwegians to acquire higher technical training—except mining engineering—was abroad. In the 1870s, there were around 40–50 Norwegian students in technical schools on the Continent.[41] At the turn of the century, there were more than 200 Norwegian students in German technical schools.[42]

Was the total number of mining engineers graduating from Norway, and abroad, enough? In the early nineteenth century, the number of mining engineer students was found to be too small, which was due to too difficult admission requirements and a long and complicated study programme, according to Grethe Authén Blom.[43] In 1910, there were still complaints about too few mining engineers, although at that point the annual number of graduates had increased considerably (see Fig. 4.1). The mining engineer Holm Holmsen wrote in *Technical Magazine* that foreigners and other types of engineers filled management and engineering positions at "a number" of mining companies due to "lack of competent mining engineers".[44] A review of the national background of the managers and directors of five large-scale mining companies confirms the use of foreigners. The French mining engineer Charles Defrance established the Vigsnes Copper Company in 1865 and became its director.[45] Sulitjelma Mining Company was established by the Swedish Consul Nils Persson in 1886. At the Swedish Alten's Copper Mines the German Fr. Schütz was director from 1897 and was replaced by the Swedish Otto Witt in 1902.[46] The Swedish Per Larsson was director of Orkla Mining Company in 1906 and in 1909 the German engineer F. Esser became its technical director. After the Norwegian mining engineer Holm Egeberg Holmsen left Orkla, the Swedish mining engineer Waldemar Carlgren was hired and the English ore-dressing engineer

[41] G. Stang, "Ble det for mange ingeniører?" in *Trondheim Ingeniørhøgskole 1912–1987 Festskrift til jubileumsfeiringen 31. oktober 1987* (Trondheim, 1987), p. 34.

[42] T. Brandt and O. Nordal, *Turbulens og Tankekraft Historien om NTNU* (Oslo, 2010), p. 90.

[43] See G. A. Blom, *Fra bergseminar til teknisk høyskole* (Oslo, 1958), pp. 124–125.

[44] *Teknisk Ukeblad* (Oslo, 1910), p. 632.

[45] F. Sæland, *Bergingeniør Emil Knudsens erindringer Inntrykk fra et bergmannsliv 1856–1897* (Kongsberg, 2005), p. 8.

[46] http://www.alta.museum.no/sider/tekst.asp?side=160 [accessed 17 March 2017].

Herrmann was recruited in 1907.[47] Furthermore, highly regarded foreign engineers were sometimes contracted for specific tasks: Anton Sophus Bachke at Røros Copper Works hired Otto Witt to elaborate a plan for rock removal at one of the mines in the late nineteenth century.[48] This use of foreign professionals may indicate a shortage of Norwegian engineers.

However, the use of foreigners seemed to decline from the turn of the century. After a couple of years, companies replaced foreigners in management positions by Norwegian engineers. In 1884, the mining engineer Olaf Aabel Corneliussen became manager at Vigsnes Copper Company.[49] In 1912, August Nachmanson took over as administrating director at Orkla Mining Company and hired mining engineer Nils Erik Lenander as technical manager. In 1904 mining engineer Emil Knudsen became president of Alten's Copper Mines and the mechanical engineer Andreas Quale was appointed as director in 1905.[50] In 1892, after having worked at Vigsnes, Corneliussen became director of the Sulitjelma Mining Company.[51] The replacement of foreigners by Norwegian engineers may be explained by the increase in graduates from the late nineteenth century. Figure 4.1 shows that by the late 1870s there were mining engineers graduating almost every year, which suggests an increased supply.

Chemists and mechanical and construction engineers were sometimes recruited to management positions, which were tasks that were normally allocated to mining engineers. In 1926, mining engineer Wolmer Marlow found that "in a number of positions where one would expect to find mining engineers, these are filled by mechanical or construction engineers".[52] Was this another sign of a lack of mining engineers? Marlow's statement seemed to reflect reality to some degree, at least from the turn

[47] Orkla Grube-Aktiebolag, *Løkken Verk En Norsk Grube Gjennom 300 År* (Trondhjem, 1954), pp. 358–364.

[48] F. Sæland, *Bergingeniør Emil Knudsens erindringer Inntrykk fra et bergmannsliv 1856–1897* (Kongsberg, 2005), p. 109.

[49] *Studenterne fra 1868* (Kristiania, 1919), pp. 65–68.

[50] http://www.alta.museum.no/sider/tekst.asp?side=160 [accessed 17 April 2017].

[51] *Studenterne fra 1868* (Kristiania, 1919), pp. 65–68.

[52] Own translation: Bergingeniørforening, *Tidsskrift for kemi og bergvæsen* (Oslo, 1926), p. 31.

of the century. Botolf Bredesen, for instance, graduated as a mechanical engineer, but was hired as a "mining engineer" at Røstvangen Mines in 1906 before he became "construction engineer" at the same company in 1908.[53] Jon Mogstad, a mechanical engineer, became "ore surveyor" at Kjøli Mines in 1906. Arne Forfang, a construction engineer, worked as "chief of ore dressing" at Sulitjelma Mining Company from 1913 and Fredrik Hagerup Jenssen, a construction engineer, worked as "ore surveyor" at Folldal Mines from 1906. Some of these engineers followed careers in mining and later became managers of mining companies. Carl Nielsen, mechanical engineer, was a "mining engineer" at Folldal Mining Company from 1907 before he became manager at Stordø Pyrite Mines.[54] The mechanical technician Hans Abel Hjelm graduated from Christiania Technical School in 1884 and became manager at a mine in Ofoten.[55] Jens Westly, a chemist from Bergen Technical School, worked as a chemist at the smelting plant at Sulitjelma Mining Company from 1904 before he became "plant chief" in 1906. He was employed at Elektrokemisk Ltd. from 1916 before he became manager at Fiskå Works and technical director at Elkem Bjølvefossen Ltd. in 1928.[56]

It should be highlighted that the use of other engineers in these positions did not necessarily mean that there was a short supply of mining engineers. As seen, mining companies used increasingly complex constructions and mechanical structures, which certainly required in-depth experience in multiple knowledge fields. Engineers with an educational background in mechanical and construction engineering—combined with mining engineers—were therefore an advantage. Moreover, a closer look at the debate about mining education strongly suggests that the focus was not on a lack of mining engineers, but on the content of the instruction and teaching facilities. Holmsen's statement in *Technical Magazine* in 1910 was in fact motivated by a desire to transfer the mining

[53] O. Alstad (red.) *Trondhjemsteknikernes Matrikel Biologiske meddelelser om samtlige faste og hospiterende elever av Trondhjems Tekniske Læreanstalt 1870–1915*, (Trondhjem, 1916).

[54] O. Alstad (red.) *Trondhjemsteknikernes Matrikel Biologiske meddelelser om samtlige faste og hospiterende elever av Trondhjems Tekniske Læreanstalt 1870–1915*, (Trondhjem, 1916).

[55] *Festskrift i anledning af Kristiania Tekniske Skoles 25 aars jubilæum* (Kristiania, 1898).

[56] L. Eskedal, *BTS-matrikkelen Ingeniører uteksaminert ved Bergen Tekniske Skole 1875–1975* (Bergen, 1975).

engineering programme to the NIT, and by "lack of competent mining engineers" he referred first and foremost to scarce teaching facilities at the University rather than a shortage of engineers.

Table 4.1 shows that the number of mining engineer graduates increased dramatically around the turn of the twentieth century. In the eighteenth century, mining engineers graduated sporadically, and until 1900 there were many years without any graduates. From 1906 the number of graduates increased considerably. Both 1911 and 1912 represented peak years with 19 and 20 graduates respectively. This dramatic increase correlated with the production increase around the turn of the century, and the start-up of the large-scale electro-metallurgical industry. More, and larger, companies demanded managers and technical staff and may have contributed to the rise in mining engineering students. In the early 1920s there were fewer graduates, and in 1926 and 1927 there were no graduates, which may be explained by the recession period and the challenges many companies experienced in the 1920s due to low prices. These upturns and downturns in number of students and graduates suggest that the intake of students was led by developments and changes in the sector.

There were some variations in the annual number of graduates during the eighteenth and nineteenth centuries, but the increased number of graduates from the turn of the twentieth century suggests more available mining engineers. The exact number of workers in the mining industry per engineer at any given time is not known, but an estimate can be made. Given that the mining engineers had an average career length of 40 years, the available mining engineers in Norway—who had graduated in Norway and abroad—would be 48 in 1866 and 221 in 1940. The estimated number of workers per mining engineer varied, but remained between 35 and 92. With this estimate, mining engineers stood for between 1.4 and 2.9 per cent of the workforce (see Table 4.3).

Using the same 40 years estimate we can compare the number of mining engineers to the number of mining companies. Between 1866 and 1870 there were between 48 and 52 mining engineers, while there was a total of 142 mining companies, which means that there were not enough mining engineers to cover all companies. In 1917, the number of mining

Table 4.3 Workers per mining engineer and technician in Norway. Estimated career of 40 years

Year	Workers	Estimated number of mining engineers available	Estimated number of workers per mining engineer	Mining engineers as percentage of workforce	Estimated number of mining technicians available	Estimated number of workers per mining technician	Mining technicians as percentage of workforce
1866	3408	*48*	***71***	***1.4***	–	–	–
1870	3239	*50*	***65***	***1.5***	*8 (year 1869)*	***405***	***0.2***
1875	2978	*49*	***61***	***1.6***	*14*	***213***	***0.5***
1880	2240	*50*	***45***	***2.2***	*28*	***80***	***1.3***
1885	2383	*51*	***47***	***2.1***	*32*	***74***	***1.3***
1890	2899	*45*	***64***	***1.5***	*46*	***63***	***1.6***
1895	2015	*51*	***40***	***2.5***	*53*	***38***	***2.6***
1900	3319	*49*	***68***	***1.5***	*53*	***63***	***1.6***
1905	4768	*52*	***92***	***1.1***	*66*	***72***	***1.4***
1910	6652	*91*	***73***	***1.4***	*96*	***69***	***1.4***
1915	8917	*152*	***59***	***1.7***	*105*	***85***	***1.8***
1920	6267	*179*	***35***	***2.9***	*115*	***54***	***1.2***
1925	8427	*198*	***43***	***2.3***	*119*	***71***	***1.4***
1930	9727	*191*	***51***	***2***	*112*	***87***	***1.2***
1935	9597	*197*	***49***	***2***	*98*	***98***	***1***
1940	10,074	*221*	***46***	***2.2***	*137*	***74***	***1.4***

Based on estimations of number of mining engineer graduates and workers

Source: Statens bergskole, *Bergskolen 100 år Jubileumsberetning 1866–1966* (Trondheim, 1966); *Studenterne* (Oslo, 1855–1940); L. Eskedal, *BTS-matrikkelen Ingeniører uteksaminert ved Bergen Tekniske Skole 1875–1975* (Bergen 1975); Gløersen, *Biografiske Oplysninger*; B. Bassøe, *Ingeniørmatrikkelen Norske Sivilingeniører 1901–1955 med tillegg* (Oslo, 1961); G. Brochmann (red.) *Vi fra NTH de første 10 kull: 1910–1919* (Stavanger, 1934); O. Amundsen, *Vi fra NTH de neste 10 kull: 1920–1929* (Oslo, 1950)

companies had increased to 167 and there were 161 mining engineers, which implied almost one mining engineer per company. By 1940, the supply had increased dramatically. In 1939, there were 139 mining companies and according to the estimates there were 212 mining engineers, which suggests there were enough mining engineers to cover one or more per company.[57] The work potential of the mining engineers seemed vast, but this will be explored in the next chapter.

A total of 191 mining technicians graduated from the Kongsberg Silver Works Elementary Mining School between 1869 and 1940. There were fewer mining technicians than mining engineers, but they also increased in number, especially after the turn of the century (see Table 4.3). Between 1922 and 1937 no one graduated due to a temporary closure of the school. Using the same career estimate of 40 years, until the 1890s there were less than 50 mining technicians, but the number increased thereafter. According to the estimates, there were 119 mining technicians in 1925 and 137 in 1940. The number of workers per mining technician declined from 405 in 1870 to 74 in 1940 and they increased from 0.2 per cent of the workforce in 1870 to around 1 per cent by 1880 and 2.6 per cent in 1895. The 8 mining technicians available in 1869 were too few to cover all companies, but this changed in the early twentieth century. By 1917, there were 167 mining companies and 98 mining technicians, and in 1939–1940 there were 139 companies and 137 mining technicians, almost enough to employ one in every company.

It may very well be that the mining technicians and engineers—perhaps especially the mining engineers—were considered too many at some point. The Committee which evaluated the transfer of the mining engineering programme from the University to the NIT, expressed in 1908 that the "over-production" of mining engineers that had taken place since 1905 "was not justifiable".[58] The mining engineer Wolmer Marlow affirmed 21 years later, in 1926, that "[t]he need for mining engineers in this country is not great. It is possibly room for one or two per year".[59]

[57] Norges offisielle statistikk, *Norges Bergverksdrift* (1866–1940).

[58] Taken from A. K. Børresen, *Bergtatt: Johan H. L. Vogt – professor, rådgiver og familiemann* (Trondheim, 2011), p. 273.

[59] Bergingeniørforening, *Tidsskrift for kemi og bergvesen*, (Oslo, 1926), p. 31.

This comment was connected to the fact that the mining sector in Norway was small compared to other countries, and also compared to other industries in the country. Production augmented, and the number of mining projects increased, but the demand for mining engineers was still not very large. Even so, there was an increased interest among students to study mining engineering, which was observed in the number of applicants. In the mid-1930s, the NIT stated that the Mining Division did not have room for all of them.[60] The number of graduates kept increasing, and much more than the indicated "two mining engineers per year" graduated in the 1910s, 1920s, and 1930s.

Another indication of an overprovision of mining engineers is the fact that as many as one-third of them (127) went to other countries to work at some point during their career. Going abroad may indicate that there were too few jobs for them in Norway. Most of them (121) graduated after 1870, which was around the time the number of graduates started to increase. The two recession periods, first in the 1870s and then in the 1920s, may have led some of these mining engineers to search for work abroad. On the other hand, most of them came back to Norway after a couple of years. Only 9 of the mining engineers stayed abroad and in 34 of the cases it is not clear whether they came back or not. This suggests that lack of jobs was a temporary problem.

In sum, there was a relatively large number of mining engineers and technicians available in Norway—compared to the size of the sector—from the late nineteenth century onwards. Norway was perhaps rare in this respect. Not all mining countries had enough mining engineers. In Chile, for example, civil engineers and mining technicians often functioned as mining engineers because the number of mining engineer graduates from the University of Chile was remarkably small. From 1850 to 1940, 302 mining engineers graduated from the University of Chile, which was in itself a considerable number, but considering that the mining sector was tenfold larger than the sector in Norway, in terms of number of workers and companies, the country would have required hundreds, and from the 1880s thousands, more mining engineer graduates to match

[60] Bergingeniørforening, *Tidsskrift for kemi og bergvæsen* (Oslo, 1935), p. 99.

Norway.[61] Between 1850 and 1940 the number of mining workers per Chilean mining engineer varied between 394 and 1395 and Chilean mining engineers stood for between 0.07 and 0.2 per cent of the total mining workforce in the country.[62] In the United States, it is less clear whether there were too few mining engineers. On the one hand, David and Wright refer to the Mining and Scientific Press, which wrote in 1915 that "nearly every successful mining operation, old or new, is today in the hands of experienced technically trained men".[63] More specifically, Clark Spencer finds that 6 out of 7 mining engineers in the United States in 1921 were college-trained.[64] But, on the other hand, it is found that it was not until after the Second World War that enough mining engineers were trained to "rely on their skills".[65]

In the next chapter the career paths and travels of the Norwegian mining engineers and technicians—and what their careers meant for mining development and businesses—are explored in detail. But first we should have a look at the other graduates from technical schools, the University, and the NIT with relevant educational background who were used by the mining sector.

Other Trained Specialists

Mining engineers and technicians only represented a small share of the skills and knowledge specialisations that were used in Norwegian mining. The increased complexity of mining activities, and the gradual mechanisation and adoption of complex machinery and equipment led to an increased use of technical manuals, user's guides, and instructions, which

---

[61] See K. Ranestad, "Multinational mining companies, employment and knowledge transfer", *Business History* (2017); K. Ranestad "The mining sectors in Chile and Norway, ca. 1870–1940: the development of a knowledge gap", *Innovation and Development*, vol. 8, issue 1, 2018.

[62] K. Ranestad, "Multinational mining companies, employment and knowledge transfer", *Business History* (2017).

[63] P. David, G. and Wright, "Increasing Returns and the Genesis of American Resource Abundance", *Oxford Journals*, vol. 6, issue 2 (1997), 231.

[64] C. Spence, *Mining engineers & the American West* (New Haven and London, 1970), p. 18.

[65] K. H. Ochs, 'The rise of American mining engineers', *Technology and Culture*. Vol. 33, No. 2, (1992), pp. 282.

may indicate that miners, smelters, and other low-level workers also partly needed to calculate, read, and write as part of their daily work. The share of literate people in Norway was very high from early on compared to other European countries, which suggests that this challenge was solved easily. Fritz Hodne finds that in 1873, around 87 per cent were able to write and read and 99 per cent were able to read.[66] Other sources show that by the 1890s the estimated literacy rate was near 100 per cent.[67] According to Carlo Cipolla, more than 70 per cent of the adult population in Norway was literate by 1850 and was thus one of the countries with the highest literacy in Europe.[68]

Technical schools, the University, and the NIT—and foreign schools and universities—supplied mining companies in Norway with intermediate and higher trained technicians, chemists, construction engineers, mechanical engineers, electricians, economists, and others, in addition to mining engineers and technicians.[69] Figure 4.2 shows the recruitment of

[66] The survey was carried out in Denmark, but according to Fritz Hodne it is seems likely to believe that the conditions were similar in Norway: F. Hodne, *Norges økonomiske historie 1815–1970* (Oslo, 1981), p. 250.

[67] O'Rourke and Williamson, "Education, Globalization and Catch-Up: Scandinavia in the Swedish Mirror", *Scandinavian Economic History Review*, 43, 1995, p. 299.

[68] Denmark, Faroe Islands, Finland, Germany, Holland, Iceland, Scotland, Sweden, Switzerland: C. M. Cipolla, *Literacy and Development in the West* (Baltimore, 1969), p. 113.

[69] This section is based on all student yearbooks: Alstad, O. (red.) *Trondhjemsteknikernes Matrikel Biologiske meddelelser om samtlige faste og hospiterende elever av Trondhjems Tekniske Læreanstalt 1870–1915* (Trondhjem, 1916); *Artiummatrikler studentene* [student yearbooks] (Kristiania/Oslo, 1855–1940); Christiansen, H. O. et al. *25-års jubileumsberetning 1912–1937 Bergen Tekniske Skole Oslo Tekniske Skole Trondheim Tekniske Skole* (Norway, 1937); *Festskrift i anledning af Kristiania Tekniske Skoles 25 aars jubilæum* (Kristiania, 1898); Baggethun, R. *Horten Ingeniørhøgskole Horten tekniske skole En beretning om landets eldste tekniske skole gjennom 125 år* (Horten, 1980); Heier, S. (red.), *100 års biografisk Jubileums-festskrift, Horten Tekniske Skole 1855–1955* (Horten, 1955); KTS, *50 årsberetning om ingeniørkullet fra Kristiania Tekniske Skole 1896* (Oslo, 1946); Kristiania tekniske skole, *Ingeniørene fra KTS 1897–1947* (Oslo, 1947); *Ingeniører fra Kr.a Tekniske Skole 1897* (Kristiania, 1922); *K.T.S. ingeniørene av 1909: matrikkel utarbeidet til 25-års jubileet 1934:* Oslo, 1934; *Oslo Tekniske Skole* (Oslo, 1894); *KTS til Ingeniørkullet av 1910: 20 årsjubileum år 1910–1930* (Oslo, 1930); *Trondhjems Tekniske Mellomskoles virksomhet i de 3 første læseaar 1912–1915* (Trondhjem, 1916); *Skrift ved 50 års jubileet for ingeniørene fra K.T.S. 1894* (Oslo, 1944); L. Eskedal, *BTS-matrikkelen Ingeniører uteksaminert ved Bergen Tekniske Skole 1875–1975* (Bergen 1975); TTL, 1897–1922 (Kristiania, 1922); Trondhjem tekniske læreanstalt, *Festskrift ved Afslutningen av Trondhjems Tekniske Læreanstalts 25de Læseaar* (Tronhjem, 1895); B. Bassøe, *Ingeniørmatrikkelen Norske Sivilingeniører 1901–55 med tillegg* (Oslo, 1961); Gløersen, J. (1932): *Biografiske Oplysninger om Kandidater med eksamen fra Bergseminaret på Kongsberg*. Oslo: Den nor-

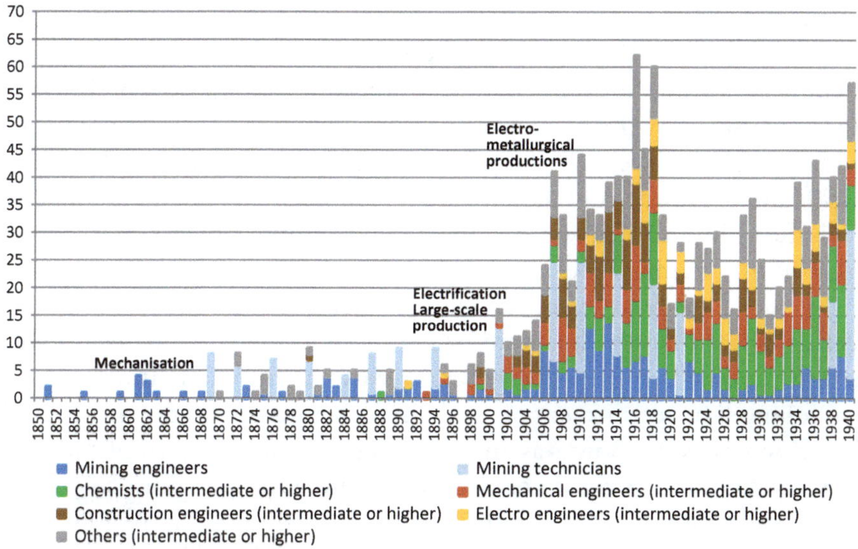

Fig. 4.2 Composition of formally trained workers recruited to mining in Norway *Year of employment.* [a]In 79 of the cases the exact year of employment is uncertain. Sources: Norges offisielle statistikk, *Norges Bergværksdrift* (Kristiania/Oslo, 1866–1940); Alstad, O. (red.) *Trondhjemsteknikernes Matrikel Biologiske meddelelser om samtlige faste og hospiterende elever av Trondhjems Tekniske Læreanstalt 1870–1915* (Trondhjem, 1916); *Artiummatrikler studentene* [student yearbooks] (Kristiania/Oslo, 1855–1940); Christiansen, H. O. et al. *25-års jubileumsberetning 1912–1937 Bergen Tekniske Skole Oslo Tekniske Skole Trondheim Tekniske Skole* (Norway, 1937); *Festskrift i anledning af Kristiania Tekniske Skoles 25 aars jubilæum* (Kristiania, 1898); Baggethun, R. *Horten Ingeniørhøgskole Horten tekniske skole En beretning om landets eldste tekniske skole gjennom 125 år* (Horten, 1980); Heier, S. (red.), *100 års biografisk Jubileums-festskrift, Horten Tekniske Skole 1855–1955* (Horten, 1955); KTS, *50 årsberetning om ingeniørkullet fra Kristiania Tekniske Skole 1896* (Oslo, 1946); Kristiania tekniske skole, *Ingeniørene fra KTS 1897–1947* (Oslo, 1947); *Ingeniører fra Kr.a Tekniske Skole 1897* (Kristiania, 1922); *K.T.S. ingeniørene av 1909: matrikkel utarbeidet til 25-års jubileet 1934:* Oslo, 1934; *Oslo Tekniske Skole* (Oslo, 1894); *KTS til Ingeniørkullet av 1910: 20 årsjubileum år 1910–1930* (Oslo, 1930); *Trondhjems Tekniske Mellomskoles virksomhet i de 3 første læseaar 1912–1915* (Trondhjem, 1916); *Skrift ved 50 års jubileet for ingeniørene fra K.T.S. 1894* (Oslo, 1944); L. Eskedal, *BTS-matrikkelen Ingeniører uteksaminert ved Bergen Tekniske Skole 1875–1975* (Bergen 1975); TTL, 1897–1922 (Kristiania, 1922); Trondhjem tekniske læreanstalt, *Festskrift ved Afslutningen av Trondhjems Tekniske Læreanstalts 25de Læseaar* (Tronhjem, 1895); B. Bassøe, *Ingeniørmatrikkelen Norske Sivilingeniører 1901–1955 med tillegg* (Oslo, 1961); Gløersen, J. (1932): *Biografiske Oplysninger om Kandidater med eksamen fra Bergseminaret på Kongsberg*. Oslo: Den norske ingeniørforening; G. Brochmann (red.) *Vi fra NTH de første 10 kull: 1910–1919* (Stavanger, 1934); O. Amundsen, *Vi fra NTH de neste 10 kull: 1920–1929* (Oslo, 1950)

all trained workers who are registered in the student yearbooks from Norway to have worked in mining between 1850 and 1940. The recruitment suggests that formally trained workers in mining became increasingly diversified by the turn of the century. Before the 1890s, a couple of mining engineers were the only formally trained workers, but from the 1890s there was a steady recruitment of workers with training and specialisations in knowledge areas other than mining, notably mechanical engineering, chemistry, electro-engineering, and construction engineering. Figure 4.2 also indicates that changes in recruitment roughly matched technological changes that were taking place in the late nineteenth and early twentieth centuries. In general terms, the mechanisation process was followed by recruitment of mechanical engineers, and the widespread adoption of electric power as an energy source from the turn of the century was followed by hiring of electricians and electro-engineers.

It should be noted that trained workers were normally recruited by companies after, or in the process, of transferring, installing, and adopting new techniques. For example, most of the electro-engineers were recruited after the turn of the century, by which time electric power was commonly used by companies in electrolysis and to provide power to mines. According to the student yearbooks, the first electro-engineer to be recruited to a mining company was Peter Kjølseth who was employed by Sulitjelma Mining Company in 1891. This was the year that the Directory and the manager decided to build an electrical power station to provide power to equipment inside and outside the mines and Kjølseth was hired to manage, and facilitate the installation of, these constructions.[70] A similar pattern seemed to be the case for chemists. The experimentation of new chemical and electro-metallurgical smelting techniques was followed by the recruitment of chemists. The large-scale electro-metallurgical iron, steel, nickel, and aluminium companies recruited most of them. In some years, in the late 1910s and the 1920s and 1930s, more chemists than mining engineers were recruited to the sector. In 1915, for example, six mining engineers and eight chemists were hired. From the turn of the century, economists, lawyers, and business administrators (categorised as "others intermediate or higher" in Fig. 4.2) were

ske ingeniørforening; G. Brochmann (red.) *Vi fra NTH de første 10 kull: 1910–1919* (Stavanger, 1934); O. Amundsen, *Vi fra NTH de neste 10 kull: 1920–1929* (Oslo, 1950).

[70] *Studenterne fra 1885* (Kristiania), 153.

also hired. They were recruited to the large-scale companies to ensure long-term planning and to facilitate rational operation and efficient management.

The recruitment of trained workers to Løkken Works (Orkla Mining Company) illustrates, on company level, the variety of specialists who were used in mining from the turn of the century and how their employment coincided with the installation and adoption of new equipment and techniques. Løkken Works started copper mining in 1654 in Meldal near Trondheim and was taken over by the Swedish Orkla Mining Company in 1904. After Orkla took over, major technological changes were implemented, which contributed to a rationalisation of operation.[71] First of all, the number of workers increased from around 10 to more than 400 in 1906, and to around 670 in 1911.[72] Second, an electric plant and advanced equipment, including pumps for drainage, air-driven drilling machines, compressors, and pumps, were installed. The decision to implement these technological changes, which led to a dramatic increase in production, was then followed by the employment of a broad set of trained workers. In the nineteenth century, only a couple of mining engineers were involved in mine organisation and ore surveying, but from the end of the nineteenth century workers with different types of education were employed.[73] The majority of the recruited trained workers were mining engineers, but also numerous electro-engineers, mechanical engineers, construction engineers, and other engineers, were hired. In 1931, the company set up a new ore-processing plant, which explains the recruitment of chemists from the late 1920s (see Table 4.4).

---

[71] Orkla Grube-Aktiebolag, *Løkken Verk En Norsk Grube Gjennom 300 År* (Trondhjem, 1954), p. 253.

[72] Norges offisielle statistikk, *Norges Bergverksdrift* (Kristiania, 1906); Norges offisielle statistikk, *Norges Bergverksdrift* (Kristiania, 1911).

[73] Henrick Christian Strøm, mining engineer from the Kongsberg Mining seminar and Freiberg Mining Academy, did surveys at the Løkken mine in the early nineteenth century; Mathias Wilhelm Sinding, mining engineer from the University of Christiania, contributed to a reorganisation of the production from copper to pyrite-copper in the mid-nineteenth century; and August Schønbek Ellefsen, mining engineer from the University of Christiania, was involved in evaluations of the operation in the late nineteenth century: Orkla Grube-Aktiebolag, *Løkken Verk En Norsk Grube Gjennom 300 År* (Trondhjem, 1954), pp. 115, 120, and 128.

Table 4.4 Recruitment of trained workers to Løkken Works (Orkla from 1904). Year of employment[a]

Year	Chemists	Construction engineers	Electro engineers	High school	Lawyers	Mechanical engineers	Military background	Mining engineers	Water, road and bridge engineers	Ship building engineers	Unknown	Total
1895–1900						1 ("manager")						**1**
1901–1905		1 ("adm. dir.")				1 ("chief engineer")		2 (1 "director")				**4**
1906–1910		3 (1 "department engineer")				2 (1 "engineer", 1 "assistant")		4 (2 "engineers")			1	**10**
1911–1915		3 (2 "engineers")	1 ("engineer")	1 (unknown)			2 (1 "transport manager")	2 ("engineers")				**9**
1916–1920		3 (1 "inspector")	4 (1 "electro-engineer", 1 "manager of electro")				1 ("accountant")	4 (1 "engineer")	1	1 ("sous chef")		**14**
1921–1925			1			1 ("assistant engineer")		2			1	**5**
1926–1930	4 (2 "chemists")		3 (1 "engineer", 1 "manager")	2 (1 "secretary", 1 "accountant")				1 ("operational manager")				**10**
1931–1935	3 (1 "chemist")		2 (1 "engineer")			6 (1 "manager of mech.", 1 "constructor", 3 "drawer")		2 ("operational engineers")			1	**14**
1936–1940	2 (unknown)	1 ("construction manager")	2 (1 "manager", 1 "engineer")	2 (1 "accountant", 1 "correspondent")	2 (1 "lawyer", 1 "secretary")	1		4 (2 "geologists", 1 "manager")	1 ("construction manager")			**15**
Total	**9**	**11**	**13**	**5**	**2**	**12**	**3**	**21**	**2**	**1**	**3**	**82**

[a]Some of the positions are unknown

Sources: *Studentene* (Oslo, 1855–1940); G. Brochmann (red.) *Vi fra NTH de første 10 kull: 1910–1919* (Stavanger, 1934); O. Amundsen, *Vi fra NTH de neste 10 kull: 1920–1929* (Oslo, 1950); L. Eskedal, *BTS-matrikkelen Ingeniører uteksaminert ved Bergen Tekniske Skole 1875–1975* (Bergen 1975); B. Bassøe, *Ingeniørmatrikkelen Norske Sivilingeniører 1901–1955 med tillegg* (Oslo, 1961); Orkla Grube-Aktiebolag, *Løkken Verk En Norsk Grube Gjennom 300 År* (Trondhjem, 1954); O. Alstad (red.) *Trondhjemsteknikernes Matrikel Biologiske meddelelser om samtlige faste og hospiterende elever av Trondhjems Tekniske Læreanstalt 1870–1915* (Trondhjem, 1916)

Table 4.4 also shows that trained workers, with some important exceptions, acquired work positions which were associated with their formal education. This reflects a general trend of the trained workers in mining; mechanical engineers were hired as "mechanics", electrical engineers were employed as "electricians", and chemists were hired as "chemists" or "metallurgists". They largely worked in the work area which their formal education indicated, which suggests that their formal knowledge specialisations were valued by the industry. It may be that a "signalling effect"[74] played a role in the recruitment process, but the general tendency of recruiting specialists to positions which correlated with their educational backgrounds strongly indicates that mining companies found their formal training to be relevant for, and directly usable in, operation. It is difficult to determine here whether formal education was a driver for technological changes in the sector, but it is clear that workers with varied formal educational backgrounds supported the development of the mining sector and the use and maintenance of the increasingly complex mining technology.

Increased diversification of the composition of trained workers is also observed in mining industries in other places in the world. English and German iron industries used natural scientists, chemists, and technicians during the nineteenth century, but in the early twentieth century a broader spectre of skilled workers and natural scientists were used, including accountants, economists, professional businessmen, and managers.[75] The North American large-scale copper projects in Chile, which started mining some of the largest copper deposits in the world in the early twentieth century with highly advanced technology, recruited a wide range of specialists and skilled workers non-stop. Correspondence between engineers of Andes Company in 1925, in the start-up phase of

[74] In Spence's approach, education is understood as part of a selection process of capable workers into suitable positions, where certain abilities indicate that the person is appropriate for a work or position. Education "signals" that the person is likely to work, but not because of the cognitive knowledge acquired through formal training: M. Spence, "Job Market Signaling", *The Quarterly Journal of Economics*, vol. 87, No. 3 (1973), 335–374.

[75] P. Musgrave, *Technical Change the Labour Force and Education* (Oxford, New York, 1967), pp. 27–135.

this corporation in Chile, indicates the broad variety of professionals that were needed. A long list of workers was to be sent from the United States to the many departments: superintendents, office engineers, foremen, field engineers, different types of construction foremen, clerks, construction mechanics, carpenters, electricians, electric constructers, draftsmen, erectors, supervisors, machinists, tank erectors, steel erectors, brick masons, pipers, and others. With regard to the machinist department, in addition to one general machinist foreman, "we should have at least six experienced machinists to install the great number of machines which go into the complete plant".[76]

Figure 4.3 shows the annual recruitment to mining of trained workers in relation to average number of workers in mining from 1866 to 1940. The total number of trained workers recruited to the sector increased dramatically in the 1900s and 1910s. In the nineteenth century, less than ten trained workers were recruited each year, while from the 1910s, more than ten, and some years more than 60, trained workers were hired. The 1910s were followed by a sharp decline in the following decade.

In 1917, the Concession Act was ratified, which coincides with this downturn. The Act was the last of the "Concession laws", which gave preference to domestic citizens and facilitated a direct state control over private and foreign companies. The Act said that the corporations' seat was to be in Norway, the majority of the boards of directors were to be Norwegian citizens and preference should be given to Norwegian workers.[77] Was it the Concession Act that led to this reduction in recruitment of local trained workers? Probably not. In the years before the ratification of the Act, companies, both domestic and multinational businesses, recruited local trained workers without being bound to any regulations, which indicates that they chose and valued them, and perhaps preferred them. Foreigners were hired, and they sometimes held managing and directing positions, but they seemed to decrease in numbers by the turn of the century as they were often

[76] Montana Historical Society Archives, Collection No. 169, *Anaconda Copper Mining Company Records,* subj. File 6.4c, folder no. 78-6, 1925–1928 staff.

[77] *Lov om erverv av vannfall mv. [industrikonsesjonsloven]* [Industrial Concession Act] (14 December 1917).

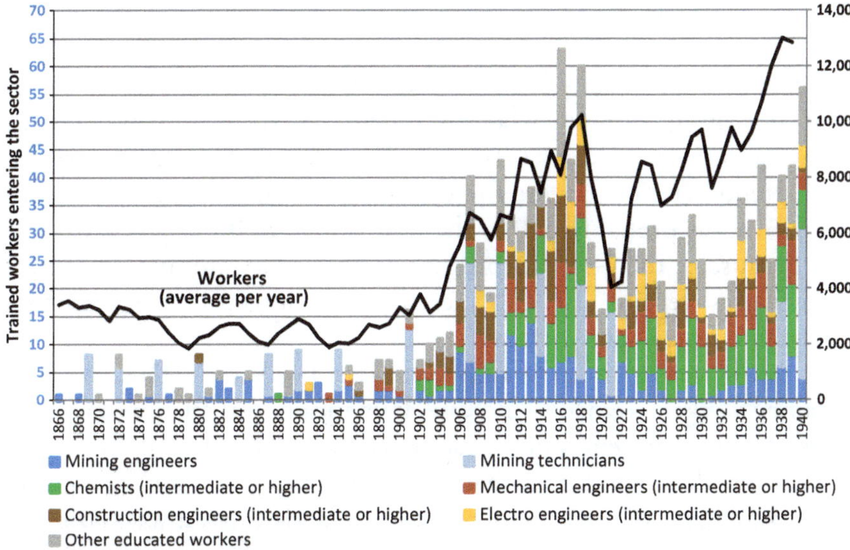

Fig. 4.3 Composition of formally trained workers recruited to the Norwegian mining sector. [a]Including people who finished high school in Norway, continued their education in Norway or abroad, and worked in Norwegian mining. [b]In 79 of the cases the exact year of employment is uncertain. Sources: Norges offisielle statistikk, *Norges Bergværksdrift* (Kristiania/Oslo, 1866–1940); Alstad, O. (red.) *Trondhjemsteknikernes Matrikel Biologiske meddelelser om samtlige faste og hospiterende elever av Trondhjems Tekniske Læreanstalt 1870–1915* (Trondhjem, 1916); Artiummatrikler studentene [student yearbooks] (Kristiania/Oslo, 1855–1940); Christiansen, H. O. et al. *25-års jubileumsberetning 1912–1937 Bergen Tekniske Skole Oslo Tekniske Skole Trondheim Tekniske Skole* (Norway, 1937); *Festskrift i anledning af Kristiania Tekniske Skoles 25 aars jubilæum* (Kristiania, 1898); Baggethun, R. *Horten Ingeniørhøgskole Horten tekniske skole En beretning om landets eldste tekniske skole gjennom 125 år* (Horten, 1980); Heier, S. (red.), *100 års biografisk Jubileums-festskrift, Horten Tekniske Skole 1855–1955* (Horten, 1955); KTS, *50 årsberetning om ingeniørkullet fra Kristiania Tekniske Skole 1896* (Oslo, 1946); Kristiania tekniske skole, *Ingeniørene fra KTS 1897–1947* (Oslo, 1947); *Ingeniører fra Kr.a Tekniske Skole 1897* (Kristiania, 1922); *K.T.S. ingeniørene av 1909: matrikkel utarbeidet til 25-års jubileet 1934:* Oslo, 1934; *Oslo Tekniske Skole* (Oslo, 1894); *KTS til Ingeniørkullet av 1910: 20 årsjubileum år 1910–1930* (Oslo, 1930); *Trondhjems Tekniske Mellomskoles virksomhet i de 3 første læseaar 1912–1915* (Trondhjem, 1916); *Skrift ved 50 års jubileet for ingeniørene fra K.T.S. 1894* (Oslo, 1944); L. Eskedal, *BTS-matrikkelen Ingeniører uteksaminert ved Bergen Tekniske Skole 1875–1975* (Bergen 1975); TTL, 1897–1922 (Kristiania, 1922); Trondhjem tekniske læreanstalt, *Festskrift ved Afslutningen av Trondhjems Tekniske Læreanstalts 25de Læseaar* (Tronhjem, 1895); B. Bassøe, *Ingeniørmatrikkelen Norske Sivilingeniører 1901–1955 med tillegg* (Oslo, 1961); Gløersen, J. (1932): *Biografiske Oplysninger om Kandidater med eksamen fra Bergseminaret på Kongsberg*. Oslo: Den norske ingeniørforening; G. Brochmann (red.) *Vi fra NTH de første 10 kull: 1910–1919* (Stavanger, 1934); O. Amundsen, *Vi fra NTH de neste 10 kull: 1920–1929* (Oslo, 1950)

replaced by Norwegians.[78] Norwegian engineers took over management of mining companies before priority to Norwegians was settled by law (see previous section). Thus, the Act formalised an already established trend. The reason for the decline in recruitment in the 1920s and 1930s was probably due to the recession period in the 1920s, and then because of the Great Depression, which made mining businesses cut expenses, reprioritise, and thereby stop the employment both of trained and untrained workers.

In spite of the decline in recruitment in the 1920s and 1930s, the share of trained workers seemed to increase dramatically. Calculations based on the year of recruitment and an estimated career of 40 years in mining suggest this, although this estimate might be an exaggeration for some of them. For engineers, mechanics, technicians, and others, it was normal to switch between sectors, and not to work in only mining during their whole career. The estimation is, thus, used as an approximation of the relationship between total workers and trained workers in the sector. Using this estimate the share of trained workers increased from 1.2 per cent in 1866 to 11.7 per cent in 1940. This drastic increase in recruitment of trained workers goes against the argument that natural resource industries, mining in this case, by definition have created few qualified jobs[79] (see Table 4.5).

The share of trained workers seemed to increase, but was the demand met? In the large-scale nickel company Falconbridge, there were a total of 275 workers and 16 director and middle manager positions in 1933,

---

[78] An exception is the aluminium industry, which developed from the early twentieth century. Aluminium companies functioned somewhat differently than other multinationals as foreign engineers were heavily involved in the development of this industry. This was found to be due to lack of knowledge in this area among Norwegian engineers: E. Storli, *Out of Norway Falls Aluminium* (Trondheim, 2010), pp. 71–73 and 148. On the other hand, there is evidence that Norwegian engineers were also heavily involved in the management and direction of this industry. First of all, Professor Harald Pedersen created a process for smelting bauxite with limestone and coke to obtain a high-quality sulphur-free pig iron and aluminium-rich slag in electric ovens. This process was the basis of the production at Norwegian Aluminium Company. Norwegian engineers and professors were initiators of some of these industrial projects and there were Norwegians on the executive boards and management and middle management positions: K. Fasting, *Norsk Aluminium gjennom 50 år* (Oslo, 1965); *Studentene fra 1904* (Oslo, 1929); *Studentene fra 1907* (Oslo, 1932); G. Brochmann, *Studentene fra 1913* (Oslo, 1938).

[79] T. Gylfason, "Natural Resources, Education and Economic Development", *European Economic Review 45* (2001).

Table 4.5 Trained workers as share of total workers. Estimated 40 years career in mining

Year	Workers (average per year)	Estimated number of trained workers in mining	Estimated trained workers as share of total workforce in mining
1866	3408	***42***	***1.2***
1870	3239	***50***	***1.5***
1875	2978	***62***	***2.1***
1880	2240	***78***	***3.5***
1885	2383	***92***	***3.9***
1890	2899	***100***	***3.4***
1895	2015	***120***	***6***
1900	3319	***140***	***4.2***
1905	4768	***188***	***3.9***
1910	6652	***330***	***5***
1915	8917	***491***	***5.5***
1920	6267	***680***	***10.9***
1925	8427	***792***	***9.4***
1930	9727	***892***	***9.2***
1935	9594	***991***	***10.3***
1940	10,074	***1178***	***11.7***

Based on estimations of number of trained workers and workers

Sources: Norges offisielle statistikk, *Norges Bergværksdrift* (Kristiania/Oslo, 1866–1940); Alstad, O. (red.) *Trondhjemsteknikernes Matrikel Biologiske meddelelser om samtlige faste og hospiterende elever av Trondhjems Tekniske Læreanstalt 1870–1915* (Trondhjem, 1916); *Artiummatrikler studentene* [student yearbooks] (Kristiania/Oslo, 1855–1940); Christiansen, H. O. et al. *25-års jubileumsberetning 1912–1937 Bergen Tekniske Skole Oslo Tekniske Skole Trondheim Tekniske Skole* (Norway, 1937); *Festskrift i anledning af Kristiania Tekniske Skoles 25 aars jubilæum* (Kristiania, 1898); Baggethun, R. *Horten Ingeniørhøgskole Horten tekniske skole En beretning om landets eldste tekniske skole gjennom 125 år* (Horten, 1980); Heier, S. (red.), *100 års biografisk Jubileums-festskrift, Horten Tekniske Skole 1855–1955* (Horten, 1955); KTS, *50 årsberetning om ingeniørkullet fra Kristiania Tekniske Skole 1896* (Oslo, 1946); Kristiania tekniske skole, *Ingeniørene fra KTS 1897–1947* (Oslo, 1947); *Ingeniører fra Kr.a Tekniske Skole 1897* (Kristiania, 1922); *K.T.S. ingeniørene av 1909: matrikkel utarbeidet til 25-års jubileet 1934:* Oslo, 1934; *Oslo Tekniske Skole* (Oslo, 1894); *KTS til Ingeniørkullet av 1910: 20 årsjubileum år 1910–1930* (Oslo, 1930); *Trondhjems Tekniske Mellomskoles virksomhet i de 3 første læseaar 1912–1915* (Trondhjem, 1916); *Skrift ved 50 års jubileet for ingeniørene fra K.T.S. 1894* (Oslo, 1944); L. Eskedal, *BTS-matrikkelen Ingeniører uteksaminert ved Bergen Tekniske Skole 1875–1975* (Bergen 1975); TTL, 1897–1922 (Kristiania, 1922); Trondhjem tekniske læreanstalt, *Festskrift ved Afslutningen av Trondhjems Tekniske Læreanstalts 25de Læseaar* (Tronhjem, 1895); B. Bassøe, *Ingeniørmatrikkelen Norske Sivilingeniører 1901–1955 med tillegg* (Oslo, 1961); Gløersen, J. (1932): *Biografiske Oplysninger om Kandidater med eksamen fra Bergseminaret på Kongsberg*. Oslo: Den norske ingeniørforening;
G. Brochmann (red.) *Vi fra NTH de første 10 kull: 1910–1919* (Stavanger, 1934); O. Amundsen, *Vi fra NTH de neste 10 kull: 1920–1929* (Oslo, 1950)

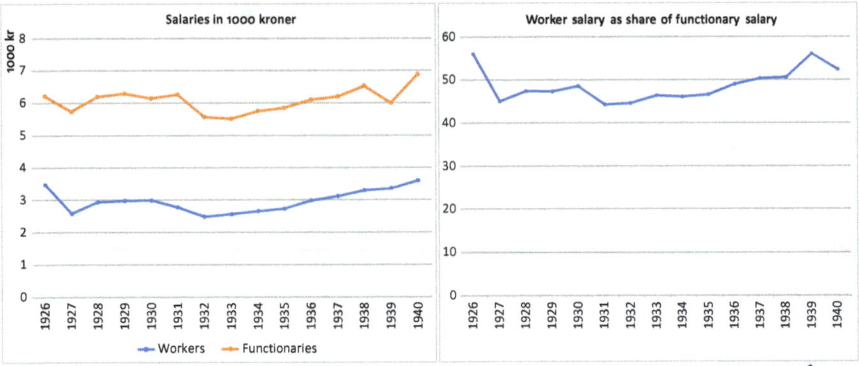

Fig. 4.4 Salaries Based on Norges offisielle statistikk, *Norges Bergverksdrift* (Oslo, 1927–1940)

totalling 5.8 per cent of the workforce.[80] The percentage of managers and middle managers probably varied from time to time, and from company to company, but it gives an indication of the number of qualified workers that were needed. In 1895, when large-scale mining production was in the early phase, trained workers stood for 6 per cent of the total mining workers, that is even more than the indicated 5.8 per cent.

An analysis of workers' and functionaries' salaries suggests that the demand of trained workers decreased relative to practical workers. From 1926, the official mining statistics include salary of "functionaries", a collective term for middle manager, manager, and director positions, which were posts that trained workers normally held. The workers' and functionaries' salaries show similar trends; they both declined in 1926 and 1927, then increased in the 1920s before they decreased slightly in the early 1930s. In the late 1930s, the functionaries' salary increased more than the workers', which suggests a larger demand in these years, but most years there was a slightly greater increase in the workers' salary than in the functionaries' salary, which suggests a large, maybe too large, supply of trained workers (see Fig. 4.4).

The ongoing debate in newspapers and technical journals confirms a large supply of engineers and technicians in Norway. The discussion was

[80] Norges offisielle statistikk, *Norges Bergverksdrift 1933* (Oslo, 1934).

not about a shortage of specialists and skilled workers—which was the case in many countries at the time—but instead about whether the technical schools, the NIT, and the University graduated too many. Many technicians and engineers went abroad to work, which was an indication that there were not enough jobs in Norway. After the creation of the three technical schools in Christiania, Bergen, and Trondheim in the 1870s, at least half of the cohorts each year left the country.[81] It was discussed whether the emigration of engineers and technicians affected industrial development negatively. With a large share of the engineer and technician graduates leaving Norway each year there might be too few left to develop industries in the country. *Technical Magazine* wrote in 1886 that

> [l]arge quantities of young skilled men, artisans and factory workers emigrate to countries where their intelligence is made useful. [...] It would be interesting to see how much technical skills that has gone to waste for our country because they have had to seek their education abroad and then remained there.[82]

Emigration of technicians and engineers also continued after the creation of the NIT in 1910. Gudmund Stang found that between 1870 and 1930 there was a gradual increase in the number of Norwegian engineering graduates leaving Norway.[83] The peak year was 1924 when more than 150 engineers from Norway went abroad.[84] In an article in 1929, Harald Pedersen, a professor at the NIT, wrote that he saw this emigration of engineers in connection with a lack of work opportunities in Norway, but he was not concerned about engineers leaving the country, and he even encouraged them to work abroad:

---

[81] G. Stang, "Ble det for mange ingeniører?" in *Trondheim Ingeniørhøgskole 1912–1987 Festskrift til jubileumsfeiringen 31. oktober 1987* (Trondheim, 1987), p. 34.

[82] Own translation: *Teknisk Ukeblad* (Kristiania, 20 May, 1886), pp. 101–102.

[83] See P-O. Grönberg, *Learning and Returning. Return Migration of Swedish Engineers from the United States, 1880–1940* (Sweden, 2003); G. Stang, "The Dispersion of Scandinavian Engineers 1870–1930 and the Concept of an Atlantic System", STS-Working Paper No. 3/89, 1992.

[84] G. Stang, "The Dispersion of Scandinavian Engineers 1870–1930 and the Concept of an Atlantic System", STS-Working Paper No. 3/89, 1992, p. 26.

> I myself find it very convenient that the engineer and architect go abroad at a young age. Some come home with useful experience, and those who remain abroad can often do great services for their country.[85]

Gudmund Stang finds that there was a massive re-emigration of engineers to Norway with vital work experience from abroad in the 1930s, which suggests that the emigration of engineers and technicians led to transfer of knowledge on a wide scale to the country.[86] The important point to be made here is that there was no general agreement that there was a lack of engineers in the country (see Chap. 5 for the return of mining engineers and knowledge transfer to Norway).

Why were there so many engineers and technicians in Norway? This was directly connected to a large number of students who enrolled in the engineering and technician programmes. Scholarships were founded from early on to technician and engineer students, which may have encouraged more students to apply to the engineering programmes. Funds were, for example, provided to mining engineer students from early on. From 1814, when the mining engineering programme was transferred from Kongsberg to the Royal Frederick University in Christiania, a specific "scholarship for mining students" was provided to "the most disadvantaged students" to pay for rent and other expenses. More importantly, however, in addition to engineering and technical study programmes being popular among students, the basic school system developed early, and facilitated students qualified for the programmes. Basic schooling, and in most cases high school exams, were required to be accepted.[87] Campaigns to improve the reading and writing skills of the Norwegian population have roots back to the seventeenth century.[88] The Church encouraged reading through religious texts from early on and the first school law in Denmark–Norway was intro-

[85] Own translation: H. Pedersen, «Den tekniske høiskole er i kontakt med storindustrien», *Bergens Tidende* (9 April 1929).

[86] G. Stang, "The Dispersion of Scandinavian Engineers 1870–1930 and the Concept of an Atlantic System", STS-Working Paper No. 3/89, 1992, pp. 26–41.

[87] UIO, *Universitets- og skoleannaler* (Kristiania, 1834–1835), pp. 214–215.

[88] K. Bruland, "Kunnskapsinstitusjoner og skandinavisk industrialisering" in *Demokratisk konservatisme,* Engelstad, F and Sejersted, F (eds.) (Oslo, 2006), p. 271.

duced in 1731. The oldest primary school in the Nordic countries opened in Bergen in 1740 and was financed by the Cross Church. During the nineteenth century, the state gradually increased public funds to education.[89] From 1827, all children in the country between 7 and 14 years of age had to be taught reading, writing, and some calculation for at least three months a year, and in 1837, 86.4 per cent of the children in the appropriate age had formal schooling. In 1860, a law which established a school system with regular school for all for seven years was introduced.[90] This widespread school system from early on assured qualified students who could potentially enter the engineering and technician programmes.

During the nineteenth century, formal technical and engineering training became increasingly used worldwide. Technical schools and universities developed across Europe, and the United States, and graduated different types of technicians, engineers, and other specialists. There was an increasing conviction that technical education led to industrial development. Peter Lundgreen finds that

> [d]uring the late nineteenth century a general belief in education as a key to international competition joined forces with a thorough transformation of the economy in providing new job markets for engineers graduating from colleges and universities.[91]

In the case of Germany especially, technical education is considered as one of the primary causes for the country taking the leading industrial role in science-based and large-scale electrical and chemical industries from the late nineteenth century. Multiple comparative studies suggest that one of the main reasons why Germany developed so strongly during what is called the "second industrial revolution", and caught up with

[89] F. Hodne and O. H. Grytten, *Norsk økonomi i det nittende århundre* (Bergen, 2000), p. 151.

[90] F. Hodne, *Norges økonomiske historie 1815–1970* (Oslo, 1981), pp. 242–244.

[91] P. Lundgreen, "Engineering Education in Europe and the U.S.A., 1750–1930: The Rise to Dominance of School Culture and the Engineering Professions", *Annals of Science*, vol. 47, Issue 1 (1990), pp. 33–75.

Britain, was a highly developed technical education system.[92] Britain's lag from the late nineteenth century, in turn, is explained by the country's less developed education system and the scarcity of technically skilled technicians and engineers.[93] More widely, it is found that European and North American countries, with a high density of engineers during the second industrial revolution, have been more innovative and have had higher growth than countries with a lower density of engineers, exemplified with poorer and underdeveloped European and Latin American countries.[94] Norway, which had an abundance of engineers and technicians from the turn of the twentieth century and a growing economy—one of the richest in the world—fits nicely into this theory. Thus, although Norway might have produced too many engineers and technicians, which may have led some of them to leave the country, having too few of them would probably lead to more severe economic consequences. In the next chapter the use of knowledge and experience by mining engineers and technicians in work and daily tasks is explored in more detail.

Summary

Formal mining engineering started in Norway in 1757 with the Mining Seminar in Kongsberg. Later, in 1867, mining education on an intermediate technical level was established with Kongsberg Silver Works Elementary Mining School. The mining engineering programme repre-

[92] See, for example, G. Ahlström, *Engineers and Industrial Growth* (London and Canberra, 1982); R. Fox, A. Guagnini, *Education, technology, and industrial performance in Europe, 1850–1939* (Cambridge, 1993); G. W. Roderick and M. D. Stephens, *Scientific and Technical Education in Nineteenth-Century England* (Newton Abbot, 1972); P. Lundgreen, "Engineering Education in Europe and the U.S.A., 1750–1930: The Rise to Dominance of School Culture and the Engineering Professions", *Annals of Science*, vol. 47, Issue 1 (1990), p. 58.

[93] M. Sanderson, *The Missing Stratum Technical School Education in England 1900–1990s* (London, 1994), p. 145.

[94] See G. Ahlström, *Engineers and Industrial Growth* (London and Canberra, 1982); F. Valencia Caicedo and W. F. Maloney, "Engineers, Innovative Capacity and Development in the Americas", *Policy Research working paper*, no. WPS 6814. Washington, DC: World Bank Group (2014).

sented a long study (4–6 years) and was theoretically oriented and more focused on natural science principles, while the mining technician programme was shorter (2–3 years) and more emphasised on practical exercises and work. The University, and the NIT from 1914, offered the mining engineering programme, which included general natural science courses in combination with mining courses. In general terms, the courses roughly coincided with the knowledge areas on which mining was based, which in turn suggests that the formal mining education was highly useful and relevant for mining.

New courses and course modifications occurred normally *after* new technology had been adopted by mining companies. For instance, new courses in mining machinery were added to the programme after mechanised equipment had been spread across the sector and electro-engineering was introduced as a new course in the early twentieth century, which was after the shift to electric power. Further adaptations were carried out in 1914 when the mining engineering instruction was moved to the NIT, which had better teaching facilities and laboratories. Specialised courses were added, and the mining engineering programme was divided in two in 1931, with one sub-programme specialising in "mining" and another in "metallurgy". These changes responded to the diversification of the sector with new mineral and metal productions and new techniques on the market, which suggests that the study programmes adapted to changes in the sector.

A total number of 341 Norwegian mining engineers graduated—in Norway or abroad—from 1787 to 1940 and according to estimates there was a gradual decrease of workers per mining engineer and mining technician. Mining engineers represented between 1.4 and 2.9 per cent of the workforce between 1866 and 1940 and mining technicians stood for between 0.2 and 2.6 per cent. Around one-third of the Norwegian mining engineers studied partly or exclusively at foreign universities and schools. This was encouraged by the University and the NIT through provision of scholarships. Students studying abroad specialised in subjects and scientific areas, which were not taught in Norway and they acquired contacts and became acquainted with foreign sciences and technology, which was key to enable transfer of technology.

The state was active in supporting education from early on, not only mining education. It financed general schooling, the University, and technical schools and managed several travel scholarship arrangements for learning purposes. Before the 1870s, a couple of mining engineers were the only formally trained workers in mining, but from the 1890s there was a steady recruitment of workers with training and specialisations in other knowledge areas than mining. Technological changes in mining were followed by a diversification in the workers' educational background from the turn of the century, notably chemists, technicians, construction engineers, mechanical engineers, electro-engineers, and economists. The trained specialists contributed with specialised knowledge, and were used by mining companies to install, operate, maintain, and organise new electric and mechanical equipment, mining constructions, mining systems, and workers. There was a correlation between education and work positions, which strongly indicates that formal education was relevant and useful for mining companies. Trained workers did not only diversify, but also increased dramatically in numbers and share from the turn of the century. According to estimates, the share of trained workers increased from 1.2 per cent in 1866 to 11.7 per cent in 1940. This indicates that mining education—and education in engineering, scientific, and technical studies more widely—took an increasing role in Norwegian mining. The increased use of trained workers coincided with the dramatic production and productivity, from the turn of the century, of numerous metals and minerals in the sector. It is difficult to determine here whether education was a driver for technological change, but it is clear that the workers with educational backgrounds supported and facilitated the increasingly complex mining technology of the late nineteenth and early twentieth centuries.

Bibliography

Manuscript Sources

Anaconda Copper Mining Company Records, Collection No. 169
Montana Historical Society Archives

Secondary Sources (Literature)

Ahlström, G. (1982). *Engineers and Industrial Growth*. London/Canberra: CROOM HELM.

Alstad, O. (red.). (1916). *Trondhjemsteknikernes Matrikel Biologiske meddelelser om samtlige faste og hospiterende elever av Trondhjems Tekniske Læreanstalt 1870–1915*. Trondhjem: F. Bruns Boghandel.

Amundsen, O. (1950). *Vi fra NTH de neste 10 kull: 1920–1929*. Oslo: Dreyer.

Artiummatrikler studentene [student yearbooks]. (1855–1940). Kristiania/Oslo.

Baggethun, R. (1980). *Horten Ingeniørhøgskole Horten tekniske skole En beretning om landets eldste tekniske skole gjennom 125 år*. Horten: Gjengangerens trykkeri.

Bassøe, B. (1961). *Ingeniørmatrikkelen Norske Sivilingeniører 1901–55 med tillegg*. Oslo: Teknisk Ukeblad.

Berg, B. I. (1998). *Gruveteknikk ved Kongsberg Sølvverk 1623–1914*. Trondheim: Senter for teknologi og samfunn, NTNU.

Bergens Tidende. (1929). Bergen.

Bergh, T., et al. (1983). *Norge fra u-land til i-land: vekst og utviklingslinjer 1830–1980*. Oslo: Gyldendal.

Bergingeniørforening. (1926). *Tidsskrift for kemi og bergvæsen*. Oslo.

Bergingeniørforening. (1932). *Tidsskrift for kemi og bergvæsen*. Oslo.

Bergingeniørforening. (1935). *Tidsskrift for kemi og bergvæsen*. Oslo.

Blom, G. A. (1958). *Fra bergseminar til teknisk høyskole*. Oslo: Norsk teknisk museum.

Boletín de la Sociedad Nacional de Minería. (1926). Santiago.

Børresen, A. K. (2011). *Bergtatt: Johan H. L. Vogt – professor, rådgiver og familiemann*. Trondheim: Tapir akademisk forlag.

Brandt, T., & Nordal, O. (2010). *Turbulens og Tankekraft Historien om NTNU*. Oslo: Pax Forlag.

Brochmann, G (red.). (1934). *Vi fra NTH de første 10 kull: 1910–1919*. Stavanger: Dreyer.

Christiansen, H. O., et al. (1937). *25-års jubileumsberetning 1912–1937 Bergen Tekniske Skole Oslo Tekniske Skole Trondheim Tekniske Skole*. Norway: Norsk Teknisk Landsforbund.

Christy, S. B. (1893, August). *The Growth of American Mining Schools and Their Relation to the Mining Industry*. University of California, Chicago Meeting, Part of the International Engineering Congress.

Cipolla, C. M. (1969). *Literacy and Development in the West*. Baltimore: Penguin Books Ltd.

David, P., & Wright, G. (1997). Increasing Returns and the Genesis of American Resource Abundance. *Industrial and Corporate Change, 6*(2), 203–246.

Den tekniske høiskole i Trondhjem. (1911–1940). *Program for studieåret…* Trondhjem: Centraltrykkeriet.

Engelstad, F., & Sejersted, F. (2006). *Demokratisk konservatisme*. Oslo: Pax.

Eskedal, L. (1975). *BTS-matrikkelen Ingeniører uteksaminert ved Bergen Tekniske Skole 1875–1975*. Bergen: A.s John Grieg.

Fasting, K. (1965). *Norsk Aluminium gjennom 50 år*. Oslo.

Festskrift i anledning af Kristiania Tekniske Skoles 25 aars jubilæum. (1898). Kristiania: J. Chr. Gundersens Bogtrykkeri.

Fox, R., & Guagnini, A. (1993). *Education, Technology, and Industrial Performance in Europe, 1850–1939*. Cambridge: University of Cambridge.

Gløersen, J. (1932). *Biografiske Oplysninger om Kandidater med eksamen fra Bergseminaret på Kongsberg*. Oslo: Den norske ingeniørforening.

Grönberg, P.-O. (2003). *Learning and Returning. Return Migration of Swedish Engineers from the United States, 1880–1940*. PhD Thesis. Sweden: Umeå University.

Gylfason, T. (2001). Natural Resources, Education and Economic Development. *European Economic Review, 45*, 847–859.

Habashi, F. (2003). *Schools of Mines the Beginnings of Mining and Metallurgical Education*. Québec: Métallurgie Extractive Québec.

Hanisch, T. J., & Lange, E. (1985). *Vitenskap for industrien NTH – En høyskole i utvikling gjennom 75 år*. Oslo: Universitetsforlaget.

Heier, S. (red.). (1955). *100 års biografisk Jubileums-festskrift, Horten Tekniske Skole 1855–1955*. Horten: A/S Gjengangerens trykkeri.

Hodne, F. (1981). *Norges økonomiske historie 1815–1970*. Oslo: Cappelen.

Hodne, F., & Grytten, O. H. (2000). *Norsk økonomi i det nittende århundre*. Bergen: Fagbokforlaget.

Ingeniører fra Kr.a Tekniske Skole 1897. (1922). Kristiania: A. W. Brøggers Boktrykkeri A/S.

K.T.S. ingeniørene av 1909: matrikkel utarbeidet til 25-års jubileet 1934 (1934). Oslo.

Kristiania tekniske skole. (1947). *Ingeniørene fra KTS 1897–1947*. Oslo.

KTS. (1946). 50 *årsberetning om ingeniørkullet fra Kristiania Tekniske Skole 1896*. Oslo.

KTS til Ingeniørkullet av 1910: 20 årsjubileum år 1910–1930. (1930). Oslo.

Lov, ang. Røraas Kobberværk: [af 12. Sept. 1818, med tillæslove]: Trondhjem.

Lov om erverv av vannfall mv. [industrikonsesjonsloven] [Industrial Concession Act], 14 December 1917.

Lundgreen, P. (1990). Engineering Education in Europe and the U.S.A, 1750–1930: The Rise to Dominance of School Culture and the Engineering Professions. *Annals of Science, 47*(1), 75.

Macchiavello Varas, S. (2010). *El problema de la industria del cobre en Chile y sus proyecciones económicas y sociales*. Santiago: Imprenta Fisca de la Penitenciaria.

Musgrave, P. (1967). *Technical Change the Labour Force and Education*. Oxford/New York: Pergamon Press.

Norges offisielle statistikk. (1866–1940). *Norges Bergverksdrift*. Kristiania/Oslo: Statistisk sentralbyrå.

Norges Tekniske Høiskole. (1920–1940). *Beretning om virksomheten...* Trondhjem: G. Krogshus Boktrykkeri A/S.

O'Rourke, K. H., & Williamson, J. G. (1995). Education, Globalization and Catch-Up: Scandinavia in the Swedish Mirror. *Scandinavian Economic History Review, 43*(3), 287–309.

Ochs, K. H. (1992). The rise of American mining engineers. *Technology and Culture, 33*(2), 278–301.

Orkla Grube-Aktiebolag. (1954). *Løkken Verk En Norsk Grube Gjennom 300 År*. Trondhjem: Orkla Grube-Aktiebolag.

Ranestad, K. (2015). *The Mining Sectors in Chile and Norway*. PhD Thesis, Geneva.

Ranestad, K. (2017). Multinational Mining Companies, Employment and Knowledge Transfer. *Business History*.

Ranestad, K. (2018). The Mining Sectors in Chile and Norway, ca. 1870–1940: The Development of a Knowledge Gap. *Innovation and Development, 8*(1), 147–165.

Roderick, G. W., & Stephens, M. D. (1972). *Scientific and Technical Education in Nineteenth-Century England*. Newton Abbot: David & Charles.

Sæland, F. (2005). *Bergingeniør Emil Knudsens erindringer Inntrykk fra et bergmannsliv 1856–1897*. Kongsberg: Norsk bergverksmuseum.

Sanderson, M. (1994). *The Missing Stratum Technical School Education in England 1900–1990s*. London: Athlone Press.

Seip, J. A., & Dahl, O. (1975). *Makt og Motiv et festskrift til Jens Arup Seip*. Oslo: Gyldendal.

Skrift ved 50 års jubileet for ingeniørene fra K.T.S. 1894 (1944). Oslo: Universal-trykkeriet.

Spence, C. (1970). *Mining Engineers & the American West*. New Haven/London: Yale University Press.

Spence, M. (1973). Job Market Signaling. *The Quarterly Journal of Economics, 87*(3), 355–374.

Stang, G. (1987). Ble det for mange ingeniører? In *Trondheim Ingeniørhøgskole 1912–1987 Festskrift til jubileumsfeiringen 31. oktober 1987*. Trondheim.

Stang, G. (1992). The Dispersion of Scandinavian Engineers 1870–1930 and the Concept of an Atlantic System, STS-Working Paper No. 3/89.

Statens bergskole. (1966). *Bergskolen 100 år Jubileumsberetning 1866–1966*. Trondheim.

Storli, E. (2010). *Out of Norway Falls Aluminium*. Doctoral theses at NTNU. Trondheim: NTNU, Faculty of Humanities, Department of History and Classical Studies.

Teknisk Ukeblad. (1886). Kristiania: Den norske ingeniør- og arkitektforening og Den polytekniske forening.

Teknisk Ukeblad. (1910). Oslo: Den norske ingeniør- og arkitektforening og Den polytekniske forening.

Trondhjem tekniske læreanstalt. (1895). *Festskrift ved Afslutningen av Trondhjems Tekniske Læreanstalts 25de Læseaar*. Trondhjem: Aktietrykkeriet i Trondhjem.

Trondhjems Tekniske Mellomskoles virksomhet i de 3 første læseaar 1912–1915 (1916). Trondhjem: Waldemar Janssens boktrykkeri.

TTL. (1922). *1897–1922*. Kristiania: C. Dahls Bok & Kunsttrykkeri.

UIO. (1834–1916). *Universitets- og skoleannaler*: Kristiania.

UIO. (1872). *Det Kongelige Norske Frederiks Universitets Aarsberetning*. Christiania: Brøgger & Christie's Bogtrykkeri.

Universidad de Chile. (1890). *Anales de la Universidad de Chile*. Santiago.

Universidad de Chile. (1908). *Anales de la Universidad de Chile*. Santiago.

Valencia Caicedo, F., & Maloney, W. F. (2014). *Engineers, Innovative Capacity and Development in the Americas*. Policy Research Working Paper; No. WPS 6814. Washington, DC: World Bank Group.

Villalobos, S., et al. (1990). *Historia de la ingenieria en Chile*. Santiago: Editorial Universitaria.

Web Pages

Alta museum. (2018, March 17). http://www.alta.museum.no/sider/tekst.asp?side=160

5

Mining Companies: Domestic and Foreign Businesses

Mining companies were normally established with private Norwegian capital, yet from the early nineteenth century, some foreign capitalists—mainly English—also invested in copper mining projects. Alten or Kaafjord Copper Works, Åmdals Copper Works, Ytterøen Mine, Bøilestad, Skattemyr Copper Mines, and Varaldsø Pyrite Mines were established with English capital in the 1830s–1860s.[1]

The recession period in the late 1870s led to the shutdown of many of the old iron, copper, and nickel works.[2] This downturn was followed by a renewal of the iron, nickel, and copper industries and the initiation of large-scale production, largely with foreign capital. New copper companies were established, such as the Swedish Sulitjelma Mining Company, the British Dunderland Iron Ore Company, and the German Sydvaranger Mine Ltd.[3] In total, the average foreign ownership in mining between 1875 and 1900 was around 41 per cent and foreigners owned 5 out of 8 large mining works in 1885.[4] Foreign investments represented 80.3 per

[1] Bergingeniørforening, *Tidsskrift for kemi og bergvæsen* (Oslo, 1930), p. 60.

[2] H. Carstens, ...*Bygger i Berge* (Trondheim, 2000), p. 108.

[3] A. Stonehill, *Foreign Ownership in Norwegian Enterprises* (Oslo, 1965), pp. 48–49.

[4] G. A. Blom, *Fra bergseminar til teknisk høyskole* (Oslo, 1958), p. 147; Norges offisielle statistikk, *Norges Bergverksdrift 1894–1895* (Oslo, 1895); Norges offisielle statistikk, *Norges Bergverksdrift 1899–1900* (Oslo, 1900).

K. Ranestad, *Knowledge-Based Growth in Natural Resource Intensive Economies*, Palgrave Studies in Economic History, https://doi.org/10.1007/978-3-319-96412-6_5

cent of total investments in mining in 1909 and thus dominated the mining sector.[5]

Another important mineral, found in the archipelago Svalbard in the early twentieth century, was coal. Eight coal companies operated in 1925, the most important being the Norwegian Store Norske Spitsbergen Kulkompani A/S and Kings Bay Kullkompani A/S. Two other companies were Norwegian, two were British-owned, one was Dutch, and one was Swedish.[6]

Several copper companies—both domestic and foreign—started to produce pyrite in addition to copper in the late nineteenth century. Ytterøy Copper Works and the French Vigsnes Copper Works started pyrite production (which included copper) in the 1860s. Røros Copper Works produced pyrite from around 1900. Arthur Stonehill described the foreign pyrite corporations in the early twentieth century, which were characterised by being capital-intensive and large-scale:

> The three largest foreign investments were concentrated with the mining and export of pyrites, which were valuable for their sulphur and copper content. In 1909, Swedish-owned Sulitjelma Aktiegruber, Fauske, employed 1688 persons. This made it the second largest corporate employer. Capital stock was kr. 7 021 000. A second pyrite mining company, The Foldal Copper Sulphur Co. Ltd., Lillelevdedalen, employed 530 persons, making it the seventeenth largest corporate employer. It began originally as a small copper mine, but went over to pyrite export in 1907. In 1904, British interests had purchased the company. [...] In 1909, its capital stock was kr. 5 580 000, mostly in British hands. A third pyrite mining company, Orkla-Grube Aktiebolag, Løkken Verk, was founded in 1904 with Swedish capital supplied by the Wallenberg group. In 1909, employment was 285 persons and capital stock kr. 4 500 000, mostly in Swedish hands.[7]

The new electro-metallurgical industry was largely based on large foreign investments, notably, Swedish, British, and North American.

[5] A. Stonehill, *Foreign Ownership in Norwegian Enterprises* (Oslo, 1965), p. 32.
[6] A. Stonehill, *Foreign Ownership in Norwegian Enterprises* (Oslo, 1965), p. 52.
[7] A. Stonehill, *Foreign Ownership in Norwegian Enterprises* (Oslo, 1965), p. 38.

Foreign companies operated in aluminium production from the turn of the century.[8] Elektrokemisk, producer of ferroalloys, was established with Swedish capital in 1904.[9] Other foreign companies were Killingdal Mines (The Bede Metal and Chemical Co. Ltd., British), Dunderland Iron Ore Co. Ltd. (British), Stordø Pyrite Mines Ltd. (Belgian), Sydvaranger Ltd. (German and Norwegian), and Bjørkaasen Mines (German).[10]

Bankruptcies and write-downs halved the face value of Norwegian mining shares between 1919 and 1936, much due to fall in prices. This led Norwegian companies to open up to foreign financial help. One of the most significant foreign investments was the Canadian company Falconbridge Nickel Mines Ltd., which bought the Norwegian-owned nickel smelting works at Kristiansand in 1929. Stangfjorden Electrochemical Factories Ltd., Norway's first aluminium works, was originally founded in 1897 with Norwegian capital, but its capital stock was later purchased by the British Aluminium Company. Most of the stock of Vigeland Works Ltd. was purchased by the British Aluminium Company in 1912. Norwegian Nitride Company Ltd. was founded in 1912 with 77 per cent French capital. The only Norwegian-owned company Høyangfaldene Norwegian Aluminium Company Ltd., founded in 1915, formed a 50:50 partnership with the North American company ALCOA in 1923.[11]

Why were there so many foreign investments in the Norwegian mining sector? Geoffrey Jones writes about multinationals from a historical perspective and points out several factors which may explain why multinational companies have invested in other countries, and why they have often had the advantage over local firms. First, multinationals have often had large amounts of capital and they have been willing to take high risks.[12] They have normally been large corporations and the size of the

[8] http://snl.no/.nbl_biografi/Harald_Pedersen/utdypning; [accessed 29 March 207]; A. K. Børresen and J. T. Kobberrød (red.), *Bergingeniørutdanning i Norge gjennom 250 år* (Trondheim, 2007), p. 109.

[9] A. Stonehill, *Foreign Ownership in Norwegian Enterprises*, 1965, pp. 35–36.

[10] Norges officielle statistikk, *Fabriktællingen I Kongeriket Norge 1909* (Kristiania, 1911), p. 152.

[11] A. Stonehill, *Foreign Ownership in Norwegian Enterprises* (Oslo, 1965), pp. 37–38 and 50.

[12] G. Jones, *Multinationals and Global Capitalism* (Oxford, 2005), p. 8.

firm has increased the possibility for large-scale production. This has also been the case in natural resource industries. Mining, in particular, has been a high-risk industry and has required large investments. There have been many unpredictabilities related to such productions. The extension and composition of ores have often been difficult to learn in advance and operation techniques have changed along the way. This, in turn, has involved uncertainties regarding costs, completion time, and operation performance. As ore deposits sometimes have been located in rough terrain, challenges concerning infrastructure have been common. In addition, fluctuation of prices represents another risk factor.[13] Although the mining sector in Norway represented a relatively small part of the economy, it was capital-intensive. In 1909, mining stood for 8.8 per cent of total paid-in capital to industries and 14.9 per cent of total industry capital stock.[14] In terms of corporate capital stock, mining stood for 3.9 per cent in 1919, 3.7 per cent in 1928, and 3.4 per cent in 1936.[15] In the nineteenth and early twentieth centuries in Norway, capital accumulation was low. There is a general agreement that the large flow of foreign investments was due to a lack of a strong local business class and entrepreneurs. The country lacked a large capital accumulation and a strong bourgeoisie.[16] As for the large aluminium production that started in the early twentieth century, Espen Storli finds that Norway had energy, but the Norwegian entrepreneurs lacked capital, raw materials, and technology.[17] It may be, then, that without the large foreign investments, many of the mining projects in Norway would not have been initiated.

Second, multinationals have sometimes possessed a specific type of knowledge or technology that other companies have not had access to, or not been able to use. These advantages have included "access to superior technology, information, knowledge, and know-how", according to Geoffrey Jones. Compared to smaller companies, multinationals have

[13] G. Jones, *Multinationals and Global Capitalism* (Oxford, 2005), p. 52.

[14] A. Stonehill, *Foreign Ownership in Norwegian Enterprises* (Oslo, 1965), pp. 34 and 36.

[15] A. Stonehill, *Foreign Ownership in Norwegian Enterprises* (Oslo, 1965), pp. 44 and 47.

[16] See, for instance, F. Hodne and O. H. Grytten, *Norsk økonomi i det nittende århundre* (Bergen, 2000); P. Fuglum, *Norges historie Norge i støpeskjeen: 1884–1920* (Oslo, 1995); S. Lieberman, *The Industrialization of Norway, 1800–1920* (Oslo, 1970).

[17] E. Storli, *Out of Norway Falls Aluminium* (Trondheim, 2010), pp. 71–72.

sometimes had superior organisational structures, "superior management techniques", or "better trained or educated managers".[18] Irving Gershenberg stresses that experience in managerial know-how and knowledge of how to train and motivate managers has been, in particular, one of the technological advantages which has enabled such firms to successfully compete in other countries:

> Multinational enterprises [...] possess an appreciable supply of managerial expertise which they utilize in their overseas operations. The organization theory of foreign direct investment emphasizes that, in order to be able to compete against indigenous entrepreneurs, foreign direct investors must possess some special attribute or factor of production which is sufficient to tip the competitive scale in their favour.[19]

Taking the previous into consideration, key questions explored in this chapter are whether mining companies—both multinational and domestic—contributed to knowledge development and, if so, how they did it.

Hands-On Practice and Work Experience

Debates about whether engineering and technical educational programmes should be more "practically oriented" or "scientifically and theoretically oriented" developed early on in many countries and continue even today. The underlying question of the debate has been whether engineers and technicians have been fully trained from the day of graduation or if they have needed more practical training. The core of the matter seems to be related to different views about the role of formal technical education in developing "capable" engineers.[20]

[18] G. Jones, *Multinationals and Global Capitalism* (Oxford, 2005), p. 8.

[19] I. Gershenberg, "The Training and Spread of Managerial Know-How, A Comparative Analysis of Multinational and Other Firms in Kenya", *World Development*, Vol. 15, No. 7 (1987), p. 932.

[20] See, for example, J. Harwood, «Engineering Education between Science and Practice: Rethinking the Historiography» in *History and Technology*, vol. 22, No. 1 (2006); A. Donovan, "Education, industry, and the American university" in R. Fox (ed.) *Education, technology, and industrial performance in Europe, 1850–1939* (Cambridge, 1993); E. Crawley (ed.), *Rethinking engineering education: the CDIO approach* (New York, 2007).

In Norway, such a debate began with the Mining Seminar in Kongsberg. Articles were published in journals and newspapers by engineers, professors, and company managers, who had different opinions about which courses to include and how the programme should be organised.[21] A statement made in the 1860s which argued for more practice in the mining engineering programme was that when the students graduated, they were not able to "climb the mining ladders or not even to hold a mining lamp without burning their fingers".[22] In the same line of argument, decades later, in 1910, the director of Sulitjelma Mining Company, Holm Holmsen, specified that mining engineers who graduated from the University had "all felt what it meant to start working with an education that was not of the practical kind".[23] There seemed to be no consensus with regard to how the mining engineering programme should be organised, and the debate continued, but the University, and later the NIT, informed that even though this was a "profession-oriented" study, graduates were not ready for their work profession when they graduated. The mining engineering programme largely included instruction of scientific theories, and—although it mixed different teaching forms, lectures, practical exercises, laboratory work, and several months' work practice—graduates needed more practical experience to perform successfully as mining engineers.[24] The University and the NIT expressed that they rather aimed to offer a type of instruction which would provide the "best" possible foundation and knowledge support for engineers in their future work.[25] In 1937, Professor Harald Dahl wrote that "[w]hen one speaks of the mining engineering education, the majority thinks of the study at the NIT; but it is often ignored, that this is just a first step, and that the most important part of education is still ahead". The

[21] See T. J. Hanisch and E. Lange, *Vitenskap for industrien NTH – En høyskole i utvikling gjennom 75 år* (Oslo, 1985), p. 53.

[22] Own translation. Taken from G. A. Blom, *Fra bergseminar til teknisk høyskole* (Oslo, 1958), pp. 140–142.

[23] *Teknisk Ukeblad* (Oslo, 1910), p. 632.

[24] The programme included a total of five months' practice at a mining company, normally Kongsberg Silver Works: Den Tekniske Høiskole I Trondhjem, *Program for Studieåret 1911–1912* (Trondhjem, 1912), 14–15.

[25] T. Brandt and O. Nordal, *Turbulens og Tankekraft Historien om NTNU* (Oslo, 2010), p. 110.

objective of the formal education was, he wrote, to provide general knowledge, to help develop the students' judgement, and to provide them with theoretical and technical insight. They were introduced to critical thinking and how to "reflect on the problems they later need to solve".[26] It was, thus, in the workplace that the mining engineers would practise problem-solving and learn how to use scientific theories in practice, according to him. The mining technician programme was more practically oriented than the mining engineering programme, but it was emphasised that the mining technician graduates also needed more practice to become fully skilled mine managers.[27] In analyses of mining education, and the role of technical and engineering education more generally, we should, thus, be aware that there have been limits to the knowledge that have been learnt in school settings—notably when it comes to experience-based learning and learning by doing, which are key to grasp the tacit dimension of knowledge—and that there was an awareness of such limits from early on.

Did mining engineers and technicians, then, acquire any practice after graduation? What did they do after finishing their studies? Among 341 Norwegian mining engineers who graduated between 1787 and 1940—either in Norway or abroad—307 obtained internship positions, or lower technical or assistant positions, in mining businesses or at the University or NIT the same year, or the year after graduation.[28] These work positions

[26] Bergingeniørforening, *Tidsskrift for kemi og bergvæsen*, (Oslo, 1937), pp. 157–161.

[27] Statens bergskole, *Bergskolen 100 år Jubileumsberetning 1866–1966* (Trondheim, 1966).

[28] This section is based on an analysis of all the student yearbooks: Alstad, O. (red.) *Trondhjemsteknikernes Matrikel Biologiske meddelelser om samtlige faste og hospiterende elever av Trondhjems Tekniske Læreanstalt 1870–1915* (Trondhjem, 1916); *Artiummatrikler studentene* [student yearbooks] (Kristiania/Oslo, 1855–1940); Christiansen, H. O. et al. *25-års jubileumsberetning 1912–1937 Bergen Tekniske Skole Oslo Tekniske Skole Trondheim Tekniske Skole* (Norway, 1937); *Festskrift i anledning af Kristiania Tekniske Skoles 25 aars jubilæum* (Kristiania, 1898); Baggethun, R. *Horten Ingeniørhøgskole Horten tekniske skole En beretning om landets eldste tekniske skole gjennom 125 år* (Horten, 1980); Heier, S. (red.), *100 års biografisk Jubileums-festskrift, Horten Tekniske Skole 1855–1955* (Horten, 1955); KTS, *50 årsberetning om ingeniørkullet fra Kristiania Tekniske Skole 1896* (Oslo, 1946); Kristiania tekniske skole, *Ingeniørene fra KTS 1897–1947* (Oslo, 1947); *Ingeniører fra Kr.a Tekniske Skole 1897* (Kristiania, 1922); *K.T.S. ingeniørene av 1909: matrikkel utarbeidet til 25-års jubileet 1934:* Oslo, 1934; *Oslo Tekniske Skole* (Oslo, 1894); *KTS til Ingeniørkullet av 1910: 20 årsjubileum år 1910–1930* (Oslo, 1930); *Trondhjems Tekniske Mellomskoles virksomhet i de 3 første læseaar 1912–1915* (Trondhjem, 1916); *Skrift ved 50 års jubileet for ingeniørene fra K.T.S. 1894* (Oslo, 1944); L. Eskedal, *BTS-matrikkelen Ingeniører uteksaminert ved Bergen Tekniske Skole 1875–1975* (Bergen 1975); TTL, 1897–1922 (Kristiania, 1922);

involved observing more experienced workers and working and collaborating with them, which would prepare the newly graduated mining engineers for more responsibilities and advanced and challenging tasks later on.

Most of the 341 mining engineers continued a career in mining. Only 21 mining engineers worked in other sectors, and one died young. They worked with all kinds of metals and minerals—copper, pyrite, silver, tin, coal, iron, pyrite, nickel, aluminium, and so on—but they also worked with consulting, at the NGS, the University, the NIT, and mining and technical schools. Their main work areas and work positions can be divided into three main categories: (1) engineers, middle managers, managers, and directors at mining companies; (2) consultants; and (3) teachers or researchers. Most of the mining engineers worked at mining companies, while 53 worked as consultants at some point during their career, more specifically as freelancers, academics, or as public servants. Seventeen had an academic career and became professors either at the University or the NIT. Only 15 of the mining engineers who studied for a career in mining never worked at a mining company.

An important feature which characterised the careers of the mining engineers is the long and widespread experience from mining companies. Appendix 4 lists the recruitment of mining engineers to mining companies and shows that they worked in all sorts of mining companies; big and small, domestic and multinational. Some companies—notably large-scale companies like Kongsberg Silver Works, Røros Copper Works, Big Norwegian Spitsbergen Coal Company, Bjørkåsen Mines, Falconbridge Nickel Works, Folldal Works, Løkken Works (Orkla), and Sulitjelma Mining Company—employed more mining engineers than others, but there was nevertheless a steady flow of mining engineers to all companies. This coincides with an article that was published in the *Mining Journal* in 1932, in which it was written that ever since the min-

Trondhjem tekniske læreanstalt, *Festskrift ved Afslutningen av Trondhjems Tekniske Læreanstalts 25de Læseaar* (Tronhjem, 1895); B. Bassøe, *Ingeniørmatrikkelen Norske Sivilingeniører 1901–55 med tillegg* (Oslo, 1961); Gløersen, J. (1932): *Biografiske Oplysninger om Kandidater med eksamen fra Bergseminaret på Kongsberg*. Oslo: Den norske ingeniørforening; G. Brochmann (red.) *Vi fra NTH de første 10 kull: 1910–1919* (Stavanger, 1934); O. Amundsen, *Vi fra NTH de neste 10 kull: 1920–1929* (Oslo, 1950).

ing engineering instruction was transferred to the University in 1814 there were mining engineers at "practically all the mining works".[29] The widespread recruitment of mining engineers to mining companies strongly indicates that they saw formal mining education as relevant and useful for mining operation.

Responsibilities and daily tasks of mining engineers ranged widely and were varied. Two main points can be made in terms of work positions: (1) Engineering work and middle management positions were the most common. Normal posts at the beginning of their careers were positions which involved engineering and technical tasks, work coordination, and ore surveying, such as "assistant engineer", "mine managers", "head of smelting" and "head of crushing plant", "chemist", "geologist", or "operational engineer". A total of 124 mining engineers became mine managers, operational engineers, or head engineers some time during their career and 47 mining engineers became smelting masters, smelting accountants, or metallurgists. (2) Another point is that mining engineers were normally not owners of mining companies. Only 8 mining engineers inherited or purchased a mining firm during their career. The great majority of them were not investors, but became instead directors or managers of mining companies. Nevertheless, they were often the motivators, or initiators, of large foreign-financed mining projects. For instance, the engineer Christian Thams was heavily involved in the establishment of Orkla Mining Company, the Norwegian industrialist Sam Eyde was the initiator of Elektrokemisk, and mining engineer Anton Grønningsæter contributed to the establishment of Falconbridge Nickel Works.[30] Around 30 per cent of the mining engineers, that is 97 of them, reached middle management level, while 108 became managers or operational managers; 85 of them became directors, and 15 became members of Board of Directors. In sum, around 60 per cent of them, that is 192, became managers or directors of mining businesses, NGS, or school or University divisions

[29] Bergingeniørforening, *Tidsskrift for kemi og bergvæsen*, 1932, p. 135.

[30] See T. Bergh et al. *Brytningstider: Orklas historie 1654–2004,* (Oslo, 2004); E. Petersen, *Elektrokemisk 1904–1954* (Oslo, 1953); P. Thonstad Sandvik, *Multinationals, Subsidiaries and National Business Systems* (London, 2012), pp. 35–36.

Table 5.1 Highest position acquired by the mining engineers during working career

Director/manager of a mining company/organisation	60 per cent
Middle management position at a company/organisation	30 per cent
Unknown	10 per cent

Sources: *Artiummatrikler studentene* [student yearbooks] (Kristiania/Oslo, 1855–1940); B. Bassøe, *Ingeniørmatrikkelen Norske Sivilingeniører 1901–1955 med tillegg* (Oslo, 1961)

during their professional life, normally after at least 10 years of work experience (see Table 5.1, which shows the highest position acquired during working career).

Mining engineers in other places of the world also acquired managing and administrating positions. In the United States, in the nineteenth and early twentieth centuries, mining engineers normally managed and directed mining companies. David and Wright refer to the Mining and Scientific Press, in which it was written in 1915 that "mining engineers increasingly assumed managerial and executive roles within large firms".[31] And a review of around one-third of the mining engineer graduates from University of Chile between 1850 and 1940 suggests that they also became administrators and directors of mining businesses.[32] It seems like indicates mining worldwide followed a general trend of increasing the use of engineers towards the end of the nineteenth century. Germany was a world leader in this respect. Ahlström concludes that Germany held a proportionally larger number of engineers in leading positions than France and England:

> [A] dominating proposition of the German industry, irrespectively of company size, possessed technically qualified engineers in management and leading positions towards the end of the nineteenth century and during the early twentieth century.[33]

An increasing number of formally trained engineers in leading positions suggests that their knowledge and experience were appreciated, and

[31] P. David, G. and Wright, "Increasing Returns and the Genesis of American Resource Abundance", *Oxford Journals*, vol. 6, issue 2 (1997), p. 231.

[32] K. Ranestad, "The mining sectors in Chile and Norway", thesis, Geneva, 2015.

[33] G. Ahlström, *Engineers and Industrial Growth* (London and Canberra, 1982), p. 98.

perhaps necessary, for the industry. What did formally trained mining engineers holding managing and directing positions mean for the mining sector in Norway? Directing, middle management, and management positions involved organisation work, administration, work coordination, and strategic technical and economic planning. More specifically, the widespread involvement of mining engineers in the management and direction of mining companies suggests that they contributed to the planning, administration, organisation, and strategic decision-making of the transfer and use of the new technology that was adopted in Norwegian mining from the late nineteenth and early twentieth centuries. Mining engineers thus seemed to be heavily involved in, and perhaps largely responsible for, the technological transformation that took place in this period.

The work and career paths of the mining technicians tended to be more specialised and one-sided than those of the mining engineers. They were in charge of the operation at the mines of a few of the largest mining companies and some of them might have worked at smelting plants, but there are no indications that they worked at the University and research centres as was the case with mining engineers.[34] Their specific role may be explained by the fact that companies sent workers to the Mining School in Kongsberg to be trained for lower intermediate or technical positions before returning to their old workplace after graduation. Their work tasks were understood as more practical, which explains why the mining technician programme was more practically oriented than the mining engineering programme.[35]

All the 191 mining technicians made careers in mining, and they mainly worked with copper, silver, iron, and pyrite. Mining technician graduates from all the cohorts became mine managers. They were recruited to Kongsberg Silver Works, but also by other large-scale companies with several hundred workers, namely the Swedish Sulitjelma Mining Company, Kjøli Mine, Ulefoss Iron Works, and Bossmo Mines, where they assisted with mine designing, mapping mines, and coordinating workers.[36]

---

[34] Statens bergskole, *Bergskolen 100 år Jubileumsberetning 1866–1966* (Trondheim, 1966), pp. 8–10 and 20.

[35] Statens bergskole, *Bergskolen 100 år Jubileumsberetning 1866–1966* (Trondheim, 1966), pp. 8–10 and 20.

[36] Statens bergskole, *Bergskolen 100 år Jubileumsberetning 1866–1966* (Trondheim, 1966), pp. 8–10 and 20.

Knowledge Networks

The information about the careers of all Norwegian mining engineers between 1787 and 1940 allows for a detailed analysis of the impact of their work. An important characteristic of their careers is that they switched jobs multiple times. 190—more than half—worked in three or more companies, or other mining organisations, in Norway during their career, and 128 worked in less than three.[37] Moreover, they worked in all the three main mining operational activities. Ore surveys and prospecting, designing and making mines, ore removal, ore crushing and smelting represented different stages of the mining production chain, which entailed highly different work tasks—and different types of knowledge—but mining engineers switched easily between them. For example, Richard Frederiks Stalberg graduated in 1858 and became the "mine manager administrator" at Kongsberg in 1866. In 1875, he switched job to ore processing and became a smelting master.[38] Forty-two of the mining engineers switched working fields during their career. Mining engineers also shifted easily between companies producing different types of metals and minerals. An example that illustrates this is Laurits Weidemann Meinich who worked with copper, silver, nickel, and pyrite. After graduation he became manager of Holtaalens Copper Works before he became a trainee in 1864 at Kongsberg Silver Works. From 1866 to 1877 he managed Grimeliens Copper Works, Lindvikens Pyrite Mines, Bøilestad Copper Works, and Rom Nickel Works. In 1877, he returned to Kongsberg Silver Works.[39] Only 17 mining engineers worked with only one type of metal or mineral during their whole career.

Such intense job switching among mining engineers was not common in all countries. A case study from Spain, for example, shows that technical workers between 1881 and 1936 normally made careers in only one firm and that there was hardly any mobility between mining compa-

[37] In ten of the cases the number of companies or organisations in which they worked is unknown.

[38] R. Falck-Muus, *Bergmannsutdannelsen i gamle dager, norske bergmenn til Sverige som ledd i utdannelsen* (Oslo, 1949), p. 128.

[39] J. K. Bergwitz, *Studentene fra 1856* (Kristiania, 1906), pp. 28–29.

nies.[40] And in early twentieth-century Chile, it was found that mining engineers normally dedicated their whole life to one branch, for instance coal, metal mining, or petroleum.[41] The reason why mining engineers in Norway seemed to switch more easily between companies, branches, and divisions than in other countries is not clear, but it may indicate, on the one hand, that they were more flexible and versatile and, on the other hand, that the working methods, techniques, and systems used by mining companies in Norway did not differ as much as they did in other countries.

This job switching led to knowledge transfer within the mining sector between companies, the University, the NIT, and NGS. The work of Jacob Pavel Friis is an example of how experience from one company was used to improve working techniques in another company. Around 1859, the managers at Kongsberg Silver Works evaluated the possibility of changing the drills that were used in ore removal. Steel drills were a cheaper alternative than iron drills, they were of better quality, and only one man was used instead of two. After testing, assessments, and multiple modifications the management and workers at the company were all "pleased with the new drills".[42] The following year, Friis was recruited as a mine manager at Røros Copper Works. He drew on the practical skills he had acquired at Kongsberg and replaced the iron drills there with steel drills "after much hard work".[43] The fact that job switching happened on a large scale suggests that strong networks were made between companies in Norway, which in turn suggests that transfer of know-how happened to a large degree.

Another aspect of the work of mining engineers, which led to key knowledge transfers between academia and mining industries, was job switching between the University, the NIT, and mining companies. Only 5 of the 17 mining engineers who worked in academia did not also work at a mining company at some point during their career. One example of

[40] M. A. López-Morell and M. A. Pérez de Perceval Verde, "French Civil Engineers in the Spanish Mining Industry", in *Entre technique et gestion*, M. Bertilorenzi, J.-P. Passaqui and A.-F. Garcon (eds.) (France, 2016).

[41] *Boletín de la Sociedad Nacional de Minería* (Santiago, 1928), p. 55.

[42] *Polyteknisk Tidsskrift* (Kristiania, 1864), pp. 188–193.

[43] J. P. Friis, *Direktør Jacob Pavel Friis' erindringer* (Oslo, 1944), p. 155.

this switching between academic work and industry is Alfred Getz. He was amanuensis at the Mineral Cabinet at the University between 1885 and 1889 and did geological research at the NGS. From 1889 to 1893, he worked in mining in Spain. In 1894, he bought a company called A/S Werner and in 1899 he became manager at Vigsnes Copper Works in Norway. In 1903, he became "administrating director" at Røros Copper Works. He became professor in mining at the NIT in 1912.[44] This shifting between industry and academia was part of a wider trend among natural scientists in Norway. It was highly common for researches and professors to work closely with mining and chemical industries. They collaborated with mining companies and experimented with new techniques, provided scientific expertise, and gave advice with regard to start-up strategies, surveying, and selection of new techniques.[45] Examples of such collaboration are the mining engineers Kraft Johanssen and Magne Mortenson who carried out experiments with flotation at Professor Harald Pedersen's laboratory in Trondheim, which led to the building of a flotation plant at Storvarts mine in Røros in 1927 and another plant in 1932 for the King's mine.[46] Professor Pedersen was also involved in the development of a process for the making of aluminium. He experimented with a process for smelting of bauxite with limestone and coke to obtain a high-quality sulphur-free pig iron and aluminium-rich slag in electric ovens while he worked at the NIT. Norwegian Aluminium Company based its operation on this process.[47] Laboratories at the University and the NIT were often used to conduct experiments and tests for mining companies, which led to important innovations.

A key point regarding the "job-switching tradition" in Norway is that job switching largely occurred between domestic and multinational mining companies. Around 37 per cent, that is 118, mining engineers worked at a multinational company, or more than one, sometime during their

[44] *Studenterne fra 1880* (Kristiania, 1905), pp. 110–111.

[45] For close ties between professors and industries, see A. K. Børresen and A. Wale (red.), *Vitenskap og teknologi for samfunnet? Bergfagene som kunnskapsfelt* (Trondheim, 2005); A. K. Børresen, *Bergtatt: Johan H. L. Vogt – professor, rådgiver og familiemann* (Trondheim, 2011).

[46] M. Mortenson, Utviklingslinjer i norsk oppredningsteknikk, *Særtrykk av Tiddskrift for kjemi, bergvesen og metallurgi* 2 (1949), p. 6.

[47] "Harald Pedersen": Det store norske leksikon, snl.no (7 November 2017).

Table 5.2 Multinationals, mining engineers, and potential knowledge transfer (ca. 1870–1940)

Multinational mining company	Number of recruited mining engineers		
Alten Copper Mines	3	⇨	
Bjørkaasen Mines	10	⇨	
Blika Mines	1	⇨	
Bøilestad Copper Works	1	⇨	
Dunderland Iron Mine	3	⇨	
Elektrokemisk	4	⇨	
Evje Nickel Works	10	⇨	
Killingdal Mines (Bede)	2	⇨	
Falconbridge Nickel Works	9	⇨	
Løkken Works (Orkla)	10	⇨	
Meraker Mines	8	⇨	Local companies/
Nordic Mining Company	2	⇨	organisations
Norwegian Aluminium Company	1	⇨	
Ringerike Nickel Works	8	⇨	
Røstvangen Mines	3	⇨	
Sulitjelma Mining Company	10	⇨	
Sydvaranger Mines	5	⇨	
Vigsnes Copper Works	6	⇨	
Ørnehommen Molybdenum Mines (Knaben)	5	⇨	
Åmdal Copper Works	2	⇨	

Sources: *Studentene* (1855–1940); *Norges Bergværksdrift* (1866–1940); B. Bassøe, *Ingeniørmatrikkelen Norske Sivilingeniører 1901–1955 med tillegg* (Oslo, 1961); G. Brochmann (red.) *Vi fra NTH de første 10 kull: 1910–1919* (Stavanger, 1934); O. Amundsen, *Vi fra NTH de neste 10 kull: 1920–1929* (Oslo, 1950); L. Eskedal, *BTS-matrikkelen Ingeniører uteksaminert ved Bergen Tekniske Skole 1875–1975* (Bergen 1975); O. Alstad (red.) *Trondhjemsteknikernes Matrikel Biologiske meddelelser om samtlige faste og hospiterende elever av Trondhjems Tekniske Læreanstalt 1870–1915*, (Trondhjem, 1916)

career, often as "director", "manager", or "engineer", and 103 of them were employed at local mining firms and public organisations *after* having worked at multinationals.[48] Table 5.2 lists the Norwegian mining engineers who were employed at local firms and mining organisations after having worked at multinationals.

[48] For details of their work positions, see K. Ranestad, K. Ranestad, "Multinational mining companies, employment and knowledge transfer", *Business History* (2017), pp. 7–8.

Some of the local companies recruited several mining engineers with experience from multinationals. Kongsberg Silver Works hired 11, Birtavarre Mines recruited 6, and Røros Copper Works employed 4. Public organisations also recruited mining engineers with experience from multinationals, notably NGS (7), organisations related to state management (7), the NIT (6), and Ministry of Commerce (3). In total, 68 local firms and organisations recruited one, or more, mining engineers with work experience from one, or more, multinational mining companies. This suggests that the mining multinationals in Norway were highly integrated and that strong networks developed between multinationals and domestic firms and organisations.

What did this job switching between multinational and domestic companies mean? Multinational mining corporations in Norway were characterised by having efficient organisational structures, they were technologically advanced, capital-intensive, and large-scale.[49] Taking this into consideration, learning how these companies operated would probably be very useful for the domestic mining companies. In this context, Geoffrey Jones writes that

> there can be substantial "knowledge spillovers" to the local economy from multinationals. Workers and managers trained by foreign companies may change jobs and transfer to local companies the skills and attitudes learned with foreign employers.[50]

Irving Gershenberg demonstrates two ways in which transfer of "managerial know-how" happen, both of which seemed to occur in Norway. First, Gershenberg points out that managers may be trained and promoted into more responsible positions. Trained workers were hired regularly to management and middle management positions in multinational mining companies in Norway, which suggests a high degree of promotion. Moreover, Norwegian engineers collaborated and shared experiences with foreigners, which contributed to capacity building and business develop-

[49] A. Stonehill, *Foreign Ownership in Norwegian Enterprises* (Oslo, 1965), p. 38.

[50] G. Jones, *Multinationals and Global Capitalism* (Oxford, 2005), 273.

ment.[51] Second, Gershenberg emphasises spread of know-how to other companies. He finds that "[i]t is this 'spread' effect which is generally assumed to constitute the most significant contribution of multinational firms to the development of an indigenous cadre of managers".[52] The many local workers who were hired by multinationals and thereafter switched to local firms and organisations later in their career were key to such spread of knowledge in Norway.

The work of Emil Knudsen is an illustrating example of how know-how and expertise from multinational companies were transferred to domestic companies. He became assistant director at the French Vigsnes Copper Works in 1882 and describes his experiences in his Memoires: "My title was 'ingénieur chef de service'. [...] The company was of course French. [...] It was for me a first-class practical school." He goes on to explain that he was given a lot of responsibility, but at the same time able to confer with "talented people with great experience" who encouraged him in his work.[53] He wrote that

> [t]he work in this well organised, large modern mining plant was so developed by the varied tasks that I was a completely different man than before, and had acquired practical experience not only in mining techniques, but also in smelting operation, construction techniques and statistics.[54]

After Vigsnes, he worked at the Norwegian company Røros Copper Works, first as "head engineer" and then as "administrating director" from 1895, where his practical skills in terms of techniques and organisational procedures from this multinational undoubtedly became useful.[55]

---

[51] Bergh, *Brytningstider;* Petersen, *Elektrokemisk 1904–54*; Thonstad, *Multinationals, Subsidiaries*, 35–36.

[52] I. Gershenberg, "The Training and Spread of Managerial Know-How, A Comparative Analysis of Multinational and Other Firms in Kenya", *World Development*, Vol. 15, No. 7 (1987), 931–939.

[53] F. Sæland, *Bergingeniør Emil Knudsens erindringer Inntrykk fra et bergmannsliv 1856–1897* (Kongsberg, 2005), 84.

[54] Own translation. F. Sæland, *Bergingeniør Emil Knudsens erindringer Inntrykk fra et bergmannsliv 1856–1897* (Kongsberg, 2005), 105.

[55] F. Sæland, *Bergingeniør Emil Knudsens erindringer Inntrykk fra et bergmannsliv 1856–1897* (Kongsberg, 2005), 105.

It is important to underline that knowledge transfer and absorption are highly complex and challenging processes and that absorbing knowledge, imitating and transferring technology from a competing multinational may have been particularly challenging, because the latter would have actively prevented knowledge from leaking. It is therefore found that successful adoption of technology from one company to another may possibly have happened to a lesser degree than previously thought.[56] In Norway, still, job switching happened to a large degree, which indicates extensive spread of managerial knowledge from multinational companies to local firms and organisations. This, in turn, may have contributed to the relatively equal technological level among mining companies in Norway.

Norway may have been exceptional when it comes to the substantial use of local workers by multinationals in management positions. Knowledge transfer between multinationals and local firms seemed to be uncommon other places in the world. In spite of being so crucial for economic systems, there are few examples of diffusion of managerial knowledge. Geoffrey Jones finds that multinationals investing in mining—perhaps especially in developing countries—have historically offered particularly few opportunities for the upgrading of human skills. They have in general had a tendency to hire expatriates in strategic technical and managing positions—and to handle the newest technologies and install and manage complex systems—and only used local workers for lower positions and unskilled and semi-skilled routine and physical jobs.[57] In more extreme cases, multinationals have created oligopolistic competition and "enclave" tendencies. In such cases, foreigners have managed and controlled subsidiaries and they have been more closely linked to broader global production systems than to the local economy.[58] Enclaves have similar characteristics to "clusters", but they differ in key aspects. In enclaves, foreign control has been total and knowledge transfer from multinationals to local firms has been limited.[59]

---

[56] G. Jones, *Multinationals and Global Capitalism* (Oxford, 2005), 266.

[57] G. Jones, *Multinationals and Global Capitalism* (Oxford, 2005), p. 267.

[58] D. Mackinnon and A. Cumbers, *An introduction to Economic Geography* (Edinburgh, 2007), pp. 138–139.

[59] M. Arias et al., "Large Mining Enterprises", Documentos de Trabajo en Economía, UCN, 2012, 5–6.

In Chile, for example, North American copper corporations in the early twentieth century were found to develop into enclaves, much due to an overwhelming use of foreigners.[60] Local engineers were almost entirely excluded from managing and middle management positions and there is no trace of any Chilean mining engineer working at these companies before 1940.[61] It was not until the 1960s that local workers filled strategic management and middle management positions.[62] Clusters, in contrast, are beneficial to the host region and country in that the company—among other things—uses local qualified workers, which reduces foreign control and facilitates knowledge transfer, much like they did in Norway.[63]

Why should employment by multinationals differ across in different countries? Multinationals may have preferred to use workers from their home country instead of local workers, especially in managing and middle managing positions. One reason for this may have been related to trust. Opening and operating subsidiaries in other countries have been highly challenging and headquarters may have decided to send workers with long work experience at the company to ensure control. Another reason, related to the first point, is that companies may have preferred to maintain the organisational structure.[64] A third reason for using workers from the home country is to avoid easy leakages to competitors.[65]

However, it is not clear that importing workers from other countries has always been the best economic solution for multinationals. It may in

---

[60] J. Pinto, *Chile, un caso de desarrollo frustrado* (Santiago, 1959), p. 56.

[61] See K. Ranestad, "Multinational mining companies, employment and knowledge transfer", *Business History* (2017), pp. 6–8.

[62] G. Jones, *Multinationals and Global Capitalism* (Oxford, 2005), 268. The dominant role of the multinationals in Chile was the motivation for a nationalisation policy, which began in the 1960s, called the "Chilenisation process". The process sought to integrate the foreign companies more in the domestic economy. In the early 1970s, a unanimous parliament voted in favour of a complete acquisition. The big copper companies were then completely nationalised and taken over by the state for a fee: P. Meller, "Chilean Economic Development 1880–1990", in *Diverging Paths Comparing a Century of Scandinavian and Latin American Economic Development,* M. Blomström and P. Meller (eds.) (Washington, 1991), p. 42.

[63] There is an agreement among scholars that foreign investments in Norway contributed to industrial development and fostered economic growth in the country. See, for example, E. Lange (ed.), *Teknologi i virksomhet: verkstedindustri i Norge etter 1840*, (Oslo, 1989); A. Stonehill, *Foreign Ownership in Norwegian Enterprises* (Oslo, 1965).

[64] A. Chandler, *Scale and Scope: the dynamics of industrial capitalism* (The United States, 1990).

[65] G. Jones, *Multinationals and Global Capitalism* (Oxford, 2005), 273.

fact be that they have preferred using local engineers, if they could.[66] First, it may have been cheaper to hire workers from the host country than to transport them across the globe. Second, workers from the host country have normally been more familiar with local conditions than foreign workers. When it comes to mining, knowledge of local conditions has been particularly important, as the geological ground and the composition of ore deposits vary greatly between regions and countries. It may be, then, that multinationals might even have preferred to use local workers, if possible.

In Norway, the general collection of mining multinationals in the late nineteenth and early twentieth centuries—independently of kind and type—seemed to be less bound to the use of workers from their home country. The multinationals acted similarly regardless of their country of origin, type of production, and geographical location in Norway. British, French, Swedish, North American, and Canadian multinationals, established in the south, centre, and north, all employed Norwegian trained workers in strategic director, management, and middle management positions. Even the aluminium companies, some of them "greenfield" projects—which traditionally have barely contributed to knowledge transfer—had local engineers in the Board of directors and also recruited local workers in managing and middle management positions.[67]

Similar behaviour among multinationals in Norway may indicate that local conditions actively influenced their recruitment processes. The Concession Act of 1917 determined that the corporation's seat was to be in Norway, most members of directories were to be Norwegian citizens, and that preference was to be given to Norwegian labour and materials. This Act may have encouraged the use of local workers, but more important was probably the widespread and high-quality basic and technical education system in Norway. Having the

[66] See K. Ranestad, "Multinational mining companies, employment and knowledge transfer", *Business History* (2017).

[67] Geoffrey Jones finds that the way multinationals have entered a country and the type of investment made have played a role. The willingness of a foreign company to transfer technology to the foreign affiliate may be influenced, for example, by whether that subsidiary is wholly owned or a joint venture. Greenfield investments, that is constructing new operational facilities from the ground up, have resulted in new processes and productions, but knowledge transfers from such projects are barely found: G. Jones, *Multinationals and Global Capitalism* (Oxford, 2005), 260.

right qualifications and experience were key requirements for director, management, and middle management positions in large-scale technologically complex mining companies, and higher—and technical—education facilitated qualified applicants. The ongoing supply of mining engineers, technicians, and other specialists with relevant training each year ensured that the multinationals had easy access to skilled and qualified workers for such positions (see Chap. 4 for a detailed description of the education system).

More widely, there is a general consensus that a widespread and high-quality basic and technical education system is key to ensuring recruitment of local workers by multinationals. Geoffrey Jones states that host countries can regulate multinationals and that knowledge transfer from multinationals to a large degree have depended on the "absorptive capacity" of host economies:

> [T]he impact of multinationals depends on the nature of the host economy, especially the level of human capital development. [...] The host country response will also be partly conditioned by the policies of its government. Host governments can improve the ability of their enterprises to absorb foreign technologies by investing in education and infrastructure. They can set the conditions under which foreign firms enter and operate in their country.[68]

Developing education systems is part of a process of developing local capacities, increasing absorptive capacity, and contributing to meeting the knowledge requirement of multinationals. Jones' statement is in accordance with analyses of employment and multinationals. In countries in which education systems have been weak and trained workers scarce, knowledge transfers seemed less likely to have occurred than in countries which historically have had high-quality education systems. E. Borensztein, J. De Gregorio, and J-W. Lee measure human capital level in different countries based on average years of male secondary schooling and find that foreign direct investment is an important vehicle for technology transfer in countries with "a minimum threshold of

[68] G. Jones, *Multinationals and Global Capitalism* (Oxford, 2005), 261.

human capital".[69] It is found, in particular, that high illiteracy rates and low technical educational levels have restricted knowledge transfers and employment of workers and managers by multinationals.[70] In Latin American and African countries, where education systems have been of lower quality, and where there has been a general shortage of engineers, technicians, and skilled workers, multinationals may have found it easier and more convenient to recruit workers from abroad. The engineering and technical education system in Norway was in sharp contrast to many Latin American countries, for example, where there was a chronic lack of mining engineers, engineers, and technicians in the nineteenth and twentieth centuries, and where there were far too few skilled workers for the countries' industries.[71] A limited supply of mining engineers, engineers, and technicians is found to be a key reason for the widespread use of foreigners by multinational mining companies in Chile.[72] It is therefore reasonable to believe that discrepancies in local capacity building largely explain differences in employment by multinationals across countries. In the next section, acquiring practical experience of new and up-to-date technology, which also contributed heavily to local capacity building and largely seemed to influence employment processes of multinationals, is discussed.

Travel Scholarships and Study Travels

Travelling to other countries was very common among mining engineers. Of the 341 Norwegian mining engineers (75 per cent) who graduated between 1787 and 1940, 256 went to other countries (1) to study at a foreign University or school, (2) to conduct geological surveys or to

[69] E. Borensztein, J. De Gregorio and J-W. Lee, "How does foreign direct investment affect economic growth?" *Journal of International Economics* 45 (1998), pp. 115–135.

[70] G. Jones, *Multinationals and Global Capitalism* (Oxford, 2005), p. 262.

[71] See, for example, S. Villalobos et al. *Historia de la ingenieria en Chile* (Santiago, 1990); M. Blomström and P. Meller (eds.), *Diverging Paths Comparing a Century of Scandinavian and Latin American Economic Development* (Washington, 1991); F. Valencia Caicedo and W. F. Maloney, "Engineers, Innovative Capacity and Development in the Americas", *Policy Research working paper*, no. WPS 6814. Washington, DC: World Bank Group (2014).

[72] See K. Ranestad, "Multinational mining companies, employment and knowledge transfer", *Business History* (2017).

acquire information about specific techniques, or (3) to work for a longer period at a foreign company. There was a continuous flow of mining engineers going to European countries, notably Germany, Sweden, France, and England, that is large industrial countries which had long mining traditions and which largely experimented with new mining and metallurgical methods.[73] From the 1880s, mining engineers also travelled to the United States.[74] By that time, the North American mining sector was one of the most advanced in the world, which may explain the new focus on this country.[75] Some of the mining engineers went to multiple countries during their trips.[76] Working and studying abroad involved learning and experiencing how foreign plants operated, that is their routines, techniques, production systems, and so on, and we have already seen that knowledge acquired in other countries was used directly in innovation processes by engineers after their return. Olav Steen's use of valuable working experience abroad to change working methods at Stavanger Electrical Steel Works is a good example (see Chap. 3). The fact that most mining engineers went abroad to work and study suggests that mining companies and organisations in Norway had easy access to workers with experience from abroad.

A total of 127 mining engineers worked for a longer period abroad, normally for a couple of years. Having worked abroad was highly valued by mining companies in Norway. Correspondence between Skandia Copper Works and Professor Johan Herman Lie Vogt suggests this. Vogt was contacted in January 1907, just after Skandia was established, and asked whether he could recommend a mining engineer as manager at the copper mine at Lillebotten and Narvik. One important criterion was that "[t]he man should have practiced at copper mines, and preferably

[73] See the inventions and modifications of every new machinery, equipment, furnace, converter, and so on used in global mining and metallurgy in the nineteenth and twentieth centuries in Singer et al., *A History of Technology*, 5 volumes.

[74] Thirty went on "study trip" and 46 worked in the United States.

[75] P. David, G. and Wright, "Increasing Returns and the Genesis of American Resource Abundance", *Oxford Journals*, vol. 6, issue 2 (1997).

[76] Fifty-six went on study trips to Sweden, 78 went to Germany, 31 went to Britain, and 20 went to France. Fourteen worked in Sweden, 12 in Germany, 6 in France, and 9 in Spain.

abroad".[77] This appreciation of foreign practice seemed to be widespread. Mining engineers who had worked abroad were often hired immediately after their return to Norway to take up managing and directing positions at mining companies or strategic positions at other mining organisations. Henrik Kristian Borchgrevink, for example, went to the United States in 1893 and worked as a mechanical engineer at Anaconda Copper Company in Montana. He travelled around in the United States, Canada, and Europe before returning to Norway in 1899 to become Director of Hadeland Mining Works and Norway Mining Company Ltd.[78] Another illustrating example is Harald Dahl who graduated in 1906. After having worked at mines in Norway for a couple of years, he went to France and North Africa to work in mining. In 1917, he went back to Norway and became Director of operations at Ofoten ore field. In 1923, he obtained a similar position at A/S Bjørkåsen Mines, Ballangen, also in Ofoten. From 1927 to 1930, he served as Director for Eisenerzgruve Fortuna in Harz, Germany, and in 1935, he went back to Norway and became a professor in mining at the NIT.[79] Gunnar Schjelderup went to Canada and became a chemist at British American Nickel Corp. in 1920 and then he started to work at Crucible Steel of America in 1921, before he left for Germany to work at Rheinmetall in Düsseldorf. He later returned to Norway and started to work in iron, steel, and aluminium production.[80] Of the 84 mining engineers who came back to Norway from longer work periods abroad (see Chap. 4), 62 were recruited to strategic technical and managing positions at mining companies, 1 was employed at the customs department, 1 by the University, 1 by the NIT, 1 by a technical school, 3 by the NGS of Norway, 3 by research centres, 1 by an insurance company, 1 by a machine company, 1 by a power company, and 1 by a railway company.[81] In their daily work of teaching, planning, organising, and managing mining projects, the mining engineers could then draw on their experience from abroad,

[77] Letter from Aktieselskabet Skandia Kobberverk to Professor J. H. L. Vogt, Luleå 16 jan. 1907, Privatarkiv nr. Tek 4, *Johan Herman Lie Vogt*, Eske 70.

[78] *Studenterne fra 1884* (Kristiania, 1909), p. 44.

[79] *Studentene fra 1901* (Oslo, 1926).

[80] G. Brochmann, *Studentene fra 1913* (Oslo, 1938), p. 227.

[81] In eight of the cases their positions are unknown.

which they seemed to do to a large degree. Similar trends are found in Sweden. In his analysis of Swedish engineers travelling from 1880 to 1930, Per-Olof Grönberg finds that learning experiences from abroad were highly valued by Swedish firms.[82]

Of the mining engineers who graduated between 1787 and 1940, 148 travelled abroad for shorter periods to visit mines or metallurgical plants, to study specific mining techniques on behalf of companies, to visit industrial exhibitions, or to practise at foreign mining plants (see Appendix 5 for a list of mining engineers and their travels abroad). Sixteen of these mining engineers specified in their biographies that they went on behalf of a company. With some exceptions, it was often the case that the trips were funded by companies with the purpose of going abroad to learn about a specific technique, such as "to Falun to study copper smelting techniques", "to mines in Germany to study turbines", "to Sweden to study the use of coke in smelting", "to France and Italy to study electrolytic processes", "to Germany to study water column machines for ore lifting and turbines for drainage and lifting", "to Freiberg to study smelting and amalgamation processes", "to France to study Pierre Manhès and electrolytic processes", and so on. This was largely because companies sent workers abroad with the specific task of analysing different machines, furnaces, or equipment of similar kinds. The purpose of these trips was to become acquainted with specific "types" of techniques or machines, for example, "electrolysis", "turbines for mining", or "power stations", and to select the sort that would function better at their plant in Norway. For example, Emil Knudsen went in the 1880s on behalf of Røros Copper Works to learn about "drilling machines".[83] In 1886, Anton Sophus Bachke went to Swansea and France to select the best copper-converting technique to be used at Røros.[84] Carl Casper Riiber went to Germany in 1894 on behalf of Kongsberg Silver Works to learn about electrical drilling and lifting machines and to assess which should be used at the mines

[82] P-O. Grönberg, *Learning and Returning. Return Migration of Swedish Engineers from the United States, 1880–1940* (Sweden, 2003), p. 1.

[83] F. Sæland, *Bergingeniør Emil Knudsens erindringer Inntrykk fra et bergmannsliv 1856–1897* (Kongsberg, 2005), p. 114.

[84] Bergingeniørforening, *Tidsskrift for bergvæsen*, 1916, pp. 41–42.

in Kongsberg.[85] Such trips were part of indispensable learning processes, which on the one hand contributed to practical capacity building and on the other hand facilitated the transfer of new techniques to Norway.

Many of the mining engineers went to other countries without any specific purpose other than to study or to work. An example is Emil Bertrand Münster who went to Freiberg before going to Mansfeldt, Clausthal, and other cities to study mines and plants, chemical factories, laboratories, and mineral collections.[86] Another example is Olaf Aabel Cornelissen who in 1879 went to Wales, Cornwall, Belgium, and Harzen to explore ore dressing and motorised man lifts. Later, in 1889 he visited several ore-dressing plants in Germany and transport plants in Luxembourg.[87] Julius Helverschou went to Scottish, German, and Norwegian mining districts around 1911 to study drilling and drainage methods in relation to deep drilling.[88] Fredrik Sebastian Nannestad travelled in Norway and to Germany and Sweden to learn about modern techniques and ore processing. He also went to Canada to study nickel ores.[89] Steinulf Smith-Meyer first studied coal mining in Germany in 1924 and then in 1931 he studied flotation techniques at mining works in Canada and in the United States.[90] The mechanical technician Theodor Wilhelm Holmsen even published a travelogue in *Technical Magazine* in 1898 of his trip to five Swedish ore-dressing and metallurgical plants. He had observed how operations were carried out and described in detail how the extraction process was implemented, and about the machinery and furnaces.[91] There are also examples of mining technicians going on "general" study trips. Anders Kokvoll went on a study trip to Sweden in 1939, even before graduation. Later, in the 1950s, he travelled abroad again to England, Finland, Belgium, and Germany.[92] These trips abroad enabled learning processes which were different than the knowledge learnt in

---

[85] B. I. Berg, *Gruveteknikk ved Kongsberg Sølvverk 1623–1914* (Trondheim, 1998), 421.

[86] G. A. Blom, *Fra bergseminar til teknisk høyskole* (Oslo, 1958), p. 134.

[87] *Studenterne fra 1868* (Kristiania, 1919), pp. 65–68.

[88] *Studentene fra 1904* (Oslo, 1929), pp. 170–171.

[89] *Studentene fra 1903*, pp. 294–296.

[90] *Studentene fra 1909* (Oslo, 1934), pp. 378–379.

[91] *Teknisk Ukeblad* (Kristiania, 1898), pp. 585–621.

[92] *Studentene fra 1933* (Oslo, 1958), p. 206.

school. During these trips, mining engineers and technicians acquired personal contacts, they gained experience with up-to-date converters, furnaces, mining equipment, power sources, mining business methods, and so on, and they learnt how to use, handle, and maintain such technology in practice. The knowledge they acquired was supplementary to the formal theoretical mining instruction, but not less useful. The general practical experience they obtained was part of their general education and training and it was essential to build a capable engineering cadre in Norway, and thus for the advance of mining projects in the country.

This massive travelling abroad was part of a wider "outward-looking" Nordic tradition. It was highly common for graduates from technical and engineering schools and universities in Sweden, Finland, Denmark, and Norway in the nineteenth and early twentieth centuries to go abroad to study and work.[93] Travelling and networking abroad did not seem to be a global trend, however. There seemed to be large differences between countries in this respect. There is evidence that skilled workers and engineers from European countries and the United States have had long traditions of networking and visiting other countries to learn and consult, while workers and engineers from Latin American countries, for example, did not usually go abroad.[94] A striking example is mining engineers from Chile of whom very few went abroad in the nineteenth and early twentieth centuries. Between 1923 and 1940, there is no trace of any mining engineer from Chile going abroad.[95]

Not going abroad may have had negative implications on industrial development. In Chile, for example, one of the reasons why Chilean mining engineers were not recruited to multinationals in the early twentieth century, and almost entirely excluded from management and mid-

[93] See P.-O. Grönberg, "To study or to work? A comparative perspective on Nordic engineer migration to German-speaking Europe, 1880–1930", in K. G. Hammarlund & Tomas Nilson (eds.) *Technology in Time, Space and Mind*, Högstad i Halmstad 13 (Halmstad, 2008); P-O. Grönberg, *Learning and Returning. Return Migration of Swedish Engineers from the United States, 1880–1940* (Sweden, 2003); G. Stang, "The Dispersion of Scandinavian Engineers 1870–1930 and the Concept of an Atlantic System", STS-Working Paper No. 3/89, 1992.

[94] D. Landes, *The Unbound Prometheus* (Cambridge, 1969), pp. 148–149.

[95] See K. Ranestad, "Multinational mining companies, employment and knowledge transfer", *Business History* (2017), p. 18.

dle management positions, was their lack of hands-on experience with up-to-date technology, which they would normally have gotten abroad. Internal letters between the mineral metallurgical manager Frederick Laist and Williams Wraith at Andes Copper Company from December 1925 about recruitment suggest this. Laist wrote that it was important to hire "good practical men to function as reverberatory, roaster and converter shift-foremen". They could be men from South America "who have had experience with the operation of roasters, reverberatories and converters [...], provided that their experience has really been with equipment of the size and design of that being initiated". But he was not convinced of hiring workers who only had practical experience from local companies: "I doubt whether a man accustomed to operating the little converter that Gore used to run at [the Chilean company] Gatico would be the right man for foreman in a converter plant such as ours."[96] It was mostly in the United States and other European countries that they would be faced with up-to-date technology, and most workers, technicians, and engineers, who had graduated from local mining schools or University, did not have this practical experience. This suggests that recruiters of mining multinationals in Chile did not consider the general cadre of local engineers and technicians to be qualified. One of the few Chilean engineers employed as manager of railway and water at the Andes Copper Company, Hermójenes Pizarro, had an engineering degree from the United States, which illustrates the high appreciation of foreign experience among mining businesses.[97] It is reasonable to believe that the lack of travelling and scarce experience from abroad among Chilean mining engineers—and trained workers more widely—was the main reason why multinational mining companies did not provide them with work opportunities and instead looked for workers elsewhere.

How can we explain differences in practical learning between countries? Why did so many Norwegian, and Nordic, engineers go abroad? In Norway, available funds to finance trips were fundamental to encourage students and newly graduated workers to travel abroad. In 1883, *Technical*

[96] Letter from Frederick Laist to Williams Wraith (MHSA/169/73/78-6).

[97] F. Solano Vega, *El Mineral de Potrerillos 1916–1918* (Copiapó, 1918), pp. 12–14.

Magazine strongly recommended the establishment of scholarships for study and practical learning purposes, and to enable knowledge transfer to Norway:

> The desirability of granting the country's technicians the opportunity of traveling to catch up and thoroughly study the progress of their subject, is obvious, and the funds, which may be used for this purpose, must be said to be well spent. [...] By studying the new methods and advancements in technology on-site and by discussing the details with specialists one would much faster and in a more complete way acquire the necessary thorough knowledge of the relevant matter than by being referred only to books and journals. [... This] can hardly be described clear[ly] enough.[98]

This recommendation was taken into account. From early on, public scholarships, notably "royal scholarships", were given to engineers who sought to study at a foreign university or school, or who aimed to study techniques at foreign mines.[99] There are traces of Norwegian miners travelling abroad from the seventeenth century with scholarships, and such funding was given more frequently from the 1730s.[100] Most of the mining engineers who graduated from the Kongsberg Mining Seminar obtained public travel scholarships—for long trips in Norway or abroad—either from the Mining Fund or the Fund "ad usus publicus".[101] This tradition continued into the nineteenth century. The regulation of the mining engineering programme from 1819 confirmed that

> [t]he mining engineers, who [...] have demonstrated the capability and skills in this subject, could hope to be provided public support to make journeys, not only in the realms of Norway and Sweden, but even in the foreign country, where mining has reached preferable perfection.[102]

---

[98] Own translation: *Teknisk Ukeblad* (Kristiania, 30 November, 1883), pp. 133–134.

[99] UIO, *Universitets- og skoleannaler* (Kristiania, 1834–1910).

[100] A. K. Børresen and J. T. Kobberrød (red.), *Bergingeniørutdanning i Norge gjennom 250 år* (Trondheim, 2007), p. 18.

[101] G. A. Blom, *Fra bergseminar til teknisk høyskole* (Oslo, 1958), p. 111.

[102] Own translation, R. Falck-Muus, *Bergmannsutdannelsen i gamle dager, norske bergmenn til Sverige som ledd i utdannelsen* (Oslo, 1949), p. 101.

Table 5.3 Non-exclusive list of public and private scholarships for study travels used by mining engineers

Year of establishment	Name of scholarship
Eighteenth century	Mining Fund
Eighteenth century	Fund "ad usus publicus"
From 1811? (The establishment of the University of Oslo)	Fund "scientific travels within Norway"
1811	University scholarship
Before 1830	Scholarship from Kongsberg Silver Works
Before 1830	Scholarship from Røros Copper Works
Before 1885	Scholarship from Vigsnes Works
Before 1880	Rathke scholarship
Before 1880	Summer travel scholarship
Before 1890?	Scholarship for technicians
Before 1906	Hjelmstjerne Rosencrone scholarship
1908	Mining Fund (different from the one above)
Before 1910	The American-Scandinavian Society Scholarship
1914	Orkla Fund
Before 1924	C. Sundt's scholarship
Before 1924	Norway-America Fund
Before 1924	Norwegian Military Goods Insurance-Fund
Before 1937	Norwegian Technical Institute scholarship
Before 1935	Hs Fund

Source: *Studenterne* (Oslo, 1855–1940); UIO, *Universitets- og skoleannaler* (Kristiania, 1834–1910); NTH, *Beretning om virksomheten* (1910–1940 Trondhjem)

Falck-Muus explained that "the majority [of the mining engineers from the University of Oslo] took the opportunity to travel with a scholarship to Swedish or German plants".[103] A public scholarship directed towards technical development was the "state technical scholarship" or "scholarship for technicians" and other public scholarships, university scholarships, state scholarships, and bequests were founded in the nineteenth and twentieth centuries (see Table 5.3).

[103] R. Falck-Muus, *Bergmannsutdannelsen i gamle dager, norske bergmenn til Sverige som ledd i utdannelsen* (Oslo, 1949), p. 101.

The University, and later the NIT, managed many of these funds, but mining companies also contributed heavily to foreign travels through the establishments of private funds.[104] Røros Copper Works, for example, provided funds from the late seventeenth century to miners going to Sweden.[105] Later, other companies followed. Orkla Mining Company established the "Orkla Fund" in 1914 for "skilled mining engineer graduates" with the aim of travelling within Norway and abroad.[106] Also Kongsberg Silver Works, Sulitjelma Mining Company Ltd., and Vigsnes Works had their private funds.[107] In many cases, the engineers did not specify in their biographies the type of scholarship they had availed, but wrote, for example, "scholarship" or "public scholarship", which means that there might have been more funds than the ones listed earlier. In sum, from 1850 to 1940 there are traces of 99 public and private scholarships and funds being distributed among 79 mining engineers for travels—mostly abroad—with the overall purpose of learning about mining, to acquire either general practical experience or knowledge of a specific method. There were also mining engineers travelling abroad at their own expense. One hundred and twenty five of the mining engineers did not mention any travel fund in their biographies.[108] Learning about foreign mining traditions and new techniques must have been regarded to be so important that foreign trips were undertaken even without scholarships.

Summary

Mining companies in Norway in the late nineteenth and early twentieth centuries had close ties to, and interacted and collaborated with, universities and technical schools, which in turn led to knowledge development, spread of knowledge, and innovation. This happened in multiple ways.

---

[104] UIO, *Universitets- og skoleannaler* (Kristiania, 1834–1910).

[105] R. Falck-Muus, *Bergmannsutdannelsen i gamle dager, norske bergmenn til Sverige som ledd i utdannelsen* (Oslo, 1949), pp. 114–129.

[106] NTH, *Program for studieåret 1916–1917* (Trondhjem, 1917), p. 58.

[107] Bergingeniørforening, *Tidsskrift for bergvæsen* (Oslo, 1917), p. 71.

[108] The type of scholarship and company is not always specified; thus there might be more mining engineers who acquired scholarship.

First, companies provided practical training for formally trained mining engineers and technicians in how to implement practical everyday tasks at mines and smelting plants. Most of the mining engineers acquired intern and assistant positions at mining companies and, through this, relevant and useful practice. This was important, as it was emphasised by the educational establishments that the formal training was not enough for graduates to perform in their field of work. Formal mining education—and technical and engineering education more widely—tended to be highly scientific and theoretically oriented and needed to be combined with a long practice. From these first years of practice mining engineers seemed to acquire some of the needed experience to carry out working tasks more efficiently.

Most formally trained mining engineers and technicians pursued a career in mining, which strongly indicates that they were understood by the sector as highly useful. Yet, we can go further and explore their functions and impacts by mapping their career paths, work positions, daily tasks, and responsibilities. Most, perhaps all, mining technicians became mine managers at few of the large-scale mining companies, in which daily tasks included supervision and technical work at the mines, planning and organising, and overseeing workers and equipment. Their function seemed increasingly important as the work and equipment at the mines became increasingly intricate from the late nineteenth century.

Mining engineers were more widespread around the mining sector than mining technicians. Mining engineers, in contrast to mining technicians, worked at all types of mining companies, and also research centres and educational establishments. Different types of mining companies—producing all kinds of metals and minerals, as well as small, big, local, and multinational mining corporations—recruited mining engineers. They had multiple functions in mining, notably as engineers, managers, metallurgists, entrepreneurs, and geologists, but also as consultants and professors. A general observation is that they were normally recruited to director, middle management, and management positions which entailed organisation and administration of mines, smelting plants, transport divisions, construction sites, and so on. Such positions also involved ore analyses, engineering and design work, accounting, and decision-making regarding the adoption of new working methods and equipment, which suggests that formally trained mining engineers played a key managing role in the technological transformation that took place in the late nineteenth and early twentieth centuries.

Second, another way in which mining companies—both domestic and foreign—contributed to knowledge development was by developing strong networks by enabling job switching. An interesting characteristic of the work of the mining engineers is that they switched jobs multiple times, normally 3–4 times, during their career. These strong networks enabled widespread knowledge diffusion within the sector, between multinational and local companies, as well as between universities, research centres, and companies. Such spread of know-how, and managerial and technical knowledge—especially from technologically advanced multinationals—has been essential in upgrading skills and developing specialised capacities, and they may have contributed to the technological transformation in mining being as broad as it was.

Third, mining companies contributed to knowledge development, and spread of knowledge, by funding travelling engineers. Around 75 per cent of all the mining engineers who graduated between 1787 and 1940 went to other countries (1) to study at a foreign University or school, (2) to conduct geological surveys or to acquire information about specific techniques, or (3) to work for a longer period at a foreign company. A total of 127 mining engineers worked for a longer period abroad, normally a couple of years. Such outward-looking tradition seemed to have long historical roots and was common among engineers and technicians from all the Nordic countries. More widely, European and North American skilled workers and engineers have had long traditions of networking abroad. Workers from other countries, for example Latin American countries, did not seem to have the same tradition. Learning processes after graduation differed greatly from country to country, which in turn may have had large implications for industrial development. Differences in practical learning among engineers and technicians should therefore be taken into account in analyses of knowledge development and innovation.

The Norwegian engineers, during trips abroad, acquired valuable contacts and most importantly they acquired practical know-how and in-depth understanding of technology. This was useful in their daily work after their return to Norway. They committed heavily to transferring knowledge from abroad to Norway and there is evidence that experience from abroad was directly used in transfer and adoption of new techniques. These learning processes abroad were key for the advancement of the mining sector, and mining companies—along with public organisations, the University, and the NIT which also provided such funds—facilitated them. The importance of having practical experience with relevant and useful technology, and the relevance of visits to foreign plants for capacity building and industrial development suggest that countries which did not encourage such travels and practical training might have missed out on important learning processes.

Bibliography

Manuscript Sources

Anaconda Copper Mining Company Records, Collection No. 169.
Johan Herman Lie Vogt, Privatarkiv Tek. 4.
Montana Historical Society Archives.
Trondhjem Mekaniske Verksted, Privatarkiv 310 [Trondhjem Mechanical Workshop].
Universitetsbiblioteket (Dora) [University Library in Trondheim, Dora].

Secondary Sources (Literature)

Ahlström, G. (1982). *Engineers and Industrial Growth*. London/Canberra: CROOM HELM.
Alstad, O. (red.). (1916). *Trondhjemsteknikernes Matrikel Biologiske meddelelser om samtlige faste og hospiterende elever av Trondhjems Tekniske Læreanstalt 1870–1915*. Trondhjem: F. Bruns Boghandel.
Amundsen, O. (1950). *Vi fra NTH de neste 10 kull: 1920–1929*. Oslo: Dreyer.
Arias M. et al. (2012). Large Mining Enterprises, Documentos de Trabajo en Economía, UCN.
Artiummatrikler studentene [student yearbooks] (1855–1940). Kristiania/Oslo.

Baggethun, R. (1980). *Horten Ingeniørhøgskole Horten tekniske skole En beretning om landets eldste tekniske skole gjennom 125 år*. Horten: Gjengangerens trykkeri.
Bassøe, B. (1961). *Ingeniørmatrikkelen Norske Sivilingeniører 1901–55 med tillegg*. Oslo: Teknisk Ukeblad.
Berg, B. I. (1998). *Gruveteknikk ved Kongsberg Sølvverk 1623–1914*. Trondheim: Senter for teknologi og samfunn, NTNU.
Bergh, T., et al. (2004). *Brytningstider: Orklas historie 1654–2004*. Oslo: Orion forlag.
Bergingeniørforening. (1916). *Tidsskrift for kemi og bergvæsen*. Oslo.
Bergingeniørforening. (1917). *Tidsskrift for bergvæsen*. Oslo.
Bergingeniørforening. (1930). *Tidsskrift for kemi og bergvæsen*. Oslo.
Bergingeniørforening. (1932). *Tidsskrift for kemi og bergvæsen*. Oslo.
Bergingeniørforening. (1937). *Tidsskrift for kemi og bergvæsen*. Oslo.
Bergingeniørforening. (1949). *Tidsskrift for kemi og bergvæsen*. Oslo.
Bertilorenzi, M., Passaqui, J.-P., & Garcon, A.-F. (Eds.). (2016). *Entre technique et gestion*. France: Presses des Mines.
Blom, G. A. (1958). *Fra bergseminar til teknisk høyskole*. Oslo: Norsk teknisk museum.
Blomström, M., & Meller, P. (Eds.). (1991). *Diverging Paths Comparing a Century of Scandinavian and Latin American Economic Development*. Washington, DC: Johns Hopkins.
Boletín de la Sociedad Nacional de Minería. (1928). Santiago.
Børresen, A. K. (2011). *Bergtatt: Johan H. L. Vogt – professor, rådgiver og familiemann*. Trondheim: Tapir akademisk forlag.
Børresen, A. K., & Kobberrød, J. T. (red.). (2007). *Bergingeniørutdanning i Norge gjennom 250 år*. Trondheim: Tapir akademisk forlag.
Børresen, A. K., & Wale, A. (red.). (2005). *Vitenskap og teknologi for samfunnet? Bergfagene som kunnskapsfelt*. Trondheim: Tapir akademisk forlag.
Brandt, T., & Nordal, O. (2010). *Turbulens og Tankekraft Historien om NTNU*. Oslo: Pax Forlag.
Brochmann, G (red.). (1934). *Vi fra NTH de første 10 kull: 1910–1919*. Stavanger: Dreyer.
Carstens, H. (2000). *...Bygger i Berge: en beretning om norsk bergverksdrift*. Trondheim: Norsk bergindustriforening Den norske bergingeniørforening Tapir.
Chandler, A. (1990). *Scale and Scope: The Dynamics of Industrial Capitalism*. Cambridge, MA: Harvard University Press.

Christiansen, H. O., et al. (1937). *25-års jubileumsberetning 1912–1937 Bergen Tekniske Skole Oslo Tekniske Skole Trondheim Tekniske Skole*. Norway: Norsk Teknisk Landsforbund.

Crawley, E. (Ed.). *Rethinking Engineering Education: The CDIO Approach*. New York: Springer.

David, P., & Wright, G. (1997). Increasing Returns and the Genesis of American Resource Abundance. *Industrial and Corporate Change, 6*(2), 203–246.

De Borensztein, E., Gregorio, J., & Lee, J.-W. (1998). How Does Foreign Direct Investment Affect Economic Growth? *Journal of International Economics, 45*, 115–135.

Den tekniske høiskole i Trondhjem. (1911–1940). *Program for studieåret…* Trondhjem: Centraltrykkeriet.

Eskedal, L. (1975). *BTS-matrikkelen Ingeniører uteksaminert ved Bergen Tekniske Skole 1875–1975*. Bergen: A.s John Grieg.

Falck-Muus, R. (1949). *Bergmannsutdannelsen i gamle dager, norske bergmenn til Sverige som ledd i utdannelsen*. Oslo: Norsk Teknisk Museum.

Festskrift i anledning af Kristiania Tekniske Skoles 25 aars jubilæum. (1898). Kristiania: J. Chr. Gundersens Bogtrykkeri.

Fox, R., & Guagnini, A. (1993). *Education, Technology, and Industrial Performance in Europe, 1850–1939*. Cambridge: University of Cambridge.

Friis, J. P. (1944). *Direktør Jacob Pavel Friis' erindringer*. Oslo: Cammermeyers Boghandel.

Fuglum, P. (1995). *Norges historie Norge i støpeskjeen: 1884–1920*. Oslo: Cappelen.

Gershenberg, I. (1987). The Training and Spread of Managerial Know-How, A Comparative Analysis of Multinational and Other Firms in Kenya. *World Development, 15*(7), 1987.

Gløersen, J. (1932). *Biografiske Oplysninger om Kandidater med eksamen fra Bergseminaret på Kongsberg*. Oslo: Den norske ingeniørforening.

Grönberg, P.-O. (2003). *Learning and Returning. Return Migration of Swedish Engineers from the United States, 1880–1940*. PhD Thesis. Sweden: Umeå University.

Hammarlund, K. G., & Nilson, T. (Eds.). (2008). *Technology in Time, Space and Mind*. Högstad: Halmstad 13.

Hanisch, T. J., & Lange, E. (1985). *Vitenskap for industrien NTH – En høyskole i utvikling gjennom 75 år*. Oslo: Universitetsforlaget.

Harwood, J. (2006). Engineering Education Between Science and Practice: Rethinking the Historiography. *History and Technology, 22*(1), 53–79.

Heier, S. (red.). (1955). *100 års biografisk Jubileums-festskrift, Horten Tekniske Skole 1855–1955*. Horten: A/S Gjengangerens trykkeri.

Hodne, F., & Grytten, O. H. (2000). *Norsk økonomi i det nittende århundre*. Bergen: Fagbokforlaget.

Ingeniører fra Kr.a Tekniske Skole 1897. (1922). Kristiania: A. W. Brøggers Boktrykkeri A/S.

Jones, G. (2005). *Multinationals and Global Capitalism*. Oxford: Oxford University Press.

K.T.S. ingeniørene av 1909: matrikkel utarbeidet til 25-års jubileet 1934 (1934). Oslo.

Kristiania tekniske skole. (1947). *Ingeniørene fra KTS 1897–1947*. Oslo.

KTS. (1946). *50 årsberetning om ingeniørkullet fra Kristiania Tekniske Skole 1896*. Oslo.

KTS til Ingeniørkullet av 1910: 20 årsjubileum år 1910–1930. (1930). Oslo.

Landes, D. (1969). *The Unbound Prometheus*. Cambridge: Cambridge University Press.

Lange, E. (Ed.). (1989). *Teknologi i virksomhet: verkstedindustri i Norge etter 1840*. Oslo: Ad notam forlag.

Lieberman, S. (1970). *The Industrialization of Norway, 1800–1920*. Oslo: Universitetsforlaget.

Mackinnon, D., & Cumbers, A. (2007). *An Introduction to Economic Geography*. Edinburgh: Routledge.

Norges officielle statistikk. (1911). *Fabriktællingen I Kongeriket Norge 1909*. Kristiania: Det statistiske centralbyraa.

Norges offisielle statistikk. (1895). *Norges Bergverksdrift 1894–95*. Oslo: Statistisk sentralbyrå.

Norges offisielle statistikk. (1900). *Norges Bergverksdrift 1899–1900*. Oslo: Statistisk sentralbyrå.

Oslo Tekniske Skole. (1894). *Skrift ved 50 års jubileet for ingeniørene fra K.T.S.* 1894: Oslo.

Petersen, E. (1953). *Elektrokemisk 1904–54*: Oslo.

Pinto Santa Cruz, A. (1959). *Chile, un caso de desarrollo frustrado. Santiago*: Editorial Universitaria.

Polyteknisk Tidsskrift. (1864). Kristiania.

Ranestad, K. (2015). *The Mining Sectors in Chile and Norway*. PhD Thesis, Geneva.

Ranestad, K. (2017). Multinational Mining Companies, Employment and Knowledge Transfer. *Business History*.

Sæland, F. (2005). *Bergingeniør Emil Knudsens erindringer Inntrykk fra et bergmannsliv 1856–1897*. Kongsberg: Norsk bergverksmuseum.

Singer, C. et al. (eds.). (1954–1975). *A History of Technology*, 5 volumes. Oxford: The Claredon Press.

Skrift ved 50 års jubileet for ingeniørene fra K.T.S. 1894. (1944). Oslo: Universal-trykkeriet.
Solano Vega, F. (1918). *El Mineral de Potrerillos 1916–1918*. Copiapó: Imprenta Progreso.
Stang, G. (1992). *The Dispersion of Scandinavian Engineers 1870–1930 and the Concept of an Atlantic System*. STS-Working Paper No. 3/89.
Statens bergskole. (1966). *Bergskolen 100 år Jubileumsberetning 1866–1966*. Trondheim.
Stonehill, A. (1965). *Foreign Ownership in Norwegian Enterprises*, Samfunnsøkonomiske Studier nr. 14. Oslo: Statistisk Sentralbyrå.
Storli, E. (2010). *Out of Norway Falls Aluminium*. Doctoral theses at NTNU. Trondheim: NTNU, Faculty of Humanities, Department of History and Classical Studies.
Teknisk Ukeblad. (1883). Kristiania: Den norske ingeniør- og arkitektforening og Den polytekniske forening.
Teknisk Ukeblad. (1898). Kristiania: Den norske ingeniør- og arkitektforening og Den polytekniske forening.
Teknisk Ukeblad. (1910). Oslo: Den norske ingeniør- og arkitektforening og Den polytekniske forening.
Thonstad Sandvik, P. (2012). *Multinationals, Subsidiaries and National Business Systems*. London: Pickering & Chatto (Publishers) Limited.
Trondhjem tekniske læreanstalt. (1895). *Festskrift ved Afslutningen av Trondhjems Tekniske Læreanstalts 25de Læseaar*. Trondhjem: Aktietrykkeriet i Trondhjem.
Trondhjems Tekniske Mellomskoles virksomhet i de 3 første læseaar 1912–1915. (1916). Trondhjem: Waldemar Janssens boktrykkeri.
TTL. (1922). *1897–1922*. Kristiania: C. Dahls Bok & Kunsttrykkeri.
UIO, *Universitets- og skoleannaler* [University and School Annals] (1834–1916). Kristiania.
Valencia Caicedo, F., & Maloney, W. F. (2014). *Engineers, Innovative Capacity and Development in the Americas*. Policy Research Working Paper; No. WPS 6814. Washington, DC: World Bank Group.
Villalobos, S., et al. (1990). *Historia de la ingenieria en Chile*. Santiago: Editorial Universitaria.

Web Pages

Store norske leksikon. (2018, March 29). http://snl.no/.nbl_biografi/Harald_Pedersen/utdypning

6

The Capital Goods Industry

Part of the resource curse theory is the argument that natural resource–intensive economies have weak dynamic development patterns and lack linkages to other industries and the wider economy.[1] Especially in less developed natural resource economies—notably Latin American and African countries—very few linkages between natural resource sectors and other industries have been found.[2] In Ghana, for example, the large-scale gold mining industry is not using local workshops for inputs.[3] Historical studies of the saltpetre and copper industries in Chile show similar patterns. North American and British mining companies invested in saltpetre and copper in the late nineteenth and early twentieth centuries and were hardly linked to industries in Chile.[4]

[1] For a review of literature on this subject, see B. Nelson and A. Behar, "Natural Resources, Growth and Spatially-Based Development: A View of the Literature" (Washington, DC, 2008).

[2] E. Rugraff and M. W. Hansen, *Multinational Corporations and Local Firms in Emerging Economies* (Amsterdam, 2011), p. 30.

[3] Larsen, Yankson and Fold, "Does FDI Create Linkages in Mining? The Case of Gold Mining in Ghana" in *Transnational Corporations and Development Policy*, Rugraff, Sánchez-Ancochea and Summer, 2009, pp. 248–249.

[4] See F. A. Encina, *Nuestra inferioridad Económica Sus Causas, sus Consecuencias* (Santiago, 1912); A. Pinto Santa Cruz, *Chile, un caso de desarrollo frustrado* (Santiago, 1959); J. Pinto Vallejos and L. Ortega Martínez, *Expansión minera y desarrollo industrial: un caso de crecimiento asociado (Chile*

K. Ranestad, *Knowledge-Based Growth in Natural Resource Intensive Economies*, Palgrave Studies in Economic History, https://doi.org/10.1007/978-3-319-96412-6_6

Yet, it does not mean that mining, and other natural resource industries, have created few linkages all together. There are multiple examples of mining companies creating strong connections to other industries, notably to capital goods industries, but instead of local ones they have been linked to industries and suppliers abroad. This is also the case for mining corporations in Ghana; they use inputs from other countries, and some of them internalised such activities.[5] Similarly, in the early twentieth century multinational mining companies in Latin America developed into isolated enclaves, which played out as closed towns in which machine shops, warehouses, repair shops, small foundries, and so on— and also cattle ranches and farms—supplied everything they needed.[6] This way of organising subsidiaries was common by multinationals in Latin America, especially between the two world wars.[7] And, instead of using local workshops they imported the larger share of machinery and technical services, equipment, and other inputs from abroad, mostly from Europe and the United States. Andes Copper Company in Chile even imported wood from abroad to be used in the mines in the Potrerillo plant.[8] Local mining companies in Chile also imported most of their supplies from other countries instead of using local suppliers.[9] These findings are supported by official statistics which show that importation of mining machinery and equipment to Chile increased during the nineteenth and early twentieth centuries.[10] Large-scale, late nineteenth- and early twentieth-century mining companies were often based on large investments and were technologically complex. They

1950–1914) (Santiago, 2004); A. Soto Cardenas, *Influencia británica en el salitre: origen, naturaleza y decadencia* (Santiago, 1998); M. Wilkins, *The Emergence of Multinational Enterprise: American business abroad from the colonial era to 1914* (Cambridge, 1970); L. Hiliart, *Braden Historia de una mina* (Chile, 1964).

[5] Larsen, Yankson and Fold, "Does FDI Create Linkages in Mining? The Case of Gold Mining in Ghana" in *Transnational Corporations and Development Policy*, Rugraff, Sánchez-Ancochea and Summer, 2009, pp. 248–249.

[6] M. Wilkins, *The Maturing of Multinational Enterprise: American business abroad from 1914 to 1970* (Cambridge, 1974), p. 125.

[7] G. Jones, *Multinationals and Global Capitalism* (Oxford, 2005), 274–275.

[8] M. C. Baros Mansilla, *Una historia de pioneros* (Santiago, 2006), 61.

[9] K. Ranestad, "The mining sectors in Chile and Norway", thesis, Geneva, 2015, 222–223.

[10] See K. Ranestad, "The mining sectors in Chile and Norway", thesis, Geneva, 2015, 230–321.

used all sorts of inputs; tools, engines, equipment, power stations, turbines, transport devices, lifts, and raw materials such as wood, coal, iron, and steel. Mining companies at the time could potentially create linkages to a number of different types of financial centres, input suppliers, and producers.

It is found that multinationals—normally using up-to-date technology and being highly knowledge-based—have sometimes acted as "engines of growth", much due to linkages that they could make to domestic industries.[11] Interactions and linkages between industries may create important learning processes and lead to innovation and new specialised productions. A recent report by the United Nations affirms that the

> key factor determining the benefits host countries can derive from FDI [foreign direct investments] are the linkages that foreign affiliates strike with domestically owned firms. Backward linkages from foreign affiliates to domestic firms are important channels through which intangible and tangible assets can be passed on from the former to the latter. They can contribute to the upgrading of domestic enterprises and embed foreign affiliates more firmly in host economies.[12]

But without any interaction between multinationals and domestic suppliers, vital learning and innovation processes can hardly take place. Company enclaves, especially, have impacted negatively on local economies. Chilean scholars emphasise that the large North American copper corporations in Chile were particularly damaging for the local economy.[13] More widely, industries, and countries, which have relied mostly upon the importation of foreign technology may have been cut off from

---

[11] G. Jones, *Multinationals and Global Capitalism* (Oxford, 2005), p. 262.

[12] UNCTAD, "World Investment Report 2001 Promoting Linkages" (New York and Geneva, 2001), p. xxi.

[13] For discussions about enclaves, see A. Pinto Santa Cruz, *Chile, un caso de desarrollo frustrado* (Santiago, 1959); M. J. Mamalakis, *The growth and structure of the Chilean economy* (London and New Haven, 1976); S. Macchiavello Varas, *El problema de la industria del cobre en Chile y sus proyecciones económicas y sociales* (Santiago, 2010).

key knowledge development. Nathan Rosenberg underlies this. He finds that the design and use of machinery have been vital sources of technological dynamism, flexibility, and vitality, and writes that countries which have imported equipment and machinery massively from abroad—such as Latin American and African countries—have missed such key learning processes.[14]

Considering the positive impacts that creating linkages may have on learning and knowledge development, how would such interactions between industries occur? In which conditions have they been created? A reason discussed in the literature of why companies may favour foreign sources of supply is related to control.[15] Recent reports by the United Nations reveal that corporations have often chosen foreign suppliers because they have established international supply chains and they have preferred suppliers who know their technical foundation, quality, scale, and cost needs. Companies have also chosen to produce inputs locally, that is produce them in-house instead of using established local producers, to save costs.[16] Yet, the essential point to bear in mind here is the argument that connections between business companies have occurred, for the most part, in cases in which the partners have considered the collaboration beneficial.[17] If firms have not seen the value in relating with local businesses, they would have probably looked elsewhere instead. Corporations in Latin American and African countries may have used foreign suppliers because the mechanical workshop and engineering industries there were small and outdated, and showed much lower capacity and competitiveness than industries in more developed countries.[18] In early twentieth-century Chile, the workshop industry did not supply the type of advanced and complex machinery that mining companies—or agriculture or manufacturing companies—required. The industry was

[14] N. Rosenberg, *Perspectives on Technology* (London, 1977), p. 157.

[15] G. Jones, *Multinationals and Global Capitalism* (Oxford, 2005), p. 272.

[16] UNCTAD, "World Investment Report 2001 Promoting Linkages" (New York and Geneva, 2001), p. 133.

[17] M. Arias et al., "Large Mining Enterprises", Documentos de Trabajo en Economía, UCN, 2012, p. 6.

[18] E. Rugraff and M. W. Hansen, *Multinational Corporations and Local Firms in Emerging Economies* (Amsterdam, 2011), p. 19; G. Jones, *Multinationals and Global Capitalism* (Oxford, 2005), p. 272.

marginal and by the end of the 1920s, numerous metal-mechanical companies had been liquidated or had reduced their operations.[19] In Ghana too, the same argument is used. Multinational mining companies use foreign suppliers due to limited local businesses.[20]

In more developed countries, such as the United States, Germany, and Britain, for their part, there is evidence of mining companies, and other businesses, creating linkages from early on. An important reason for this is probably the competitiveness, size, and capacity that local supply industries have shown in these countries. Engineering and technical service firms in Latin American countries have increased in number and shown increased competitiveness in recent decades, which in turn is found to be the reason for stronger linkages than in the early twentieth century. A recent study of Peru, for example, shows that mining firms use local workshops and collaborate with universities and other organisations, for power generation supply and technical services in geology, mining engineering, environmental engineering, and metallurgy.[21] In Chile, increased adaptability and modernisation of engineering firms in recent decades have intensified linkages between such firms and mining companies.[22]

Did linkages develop between the mining sector—especially multinationals—and input suppliers in Norway in the late nineteenth and early twentieth centuries? We have seen that mining in Norway became highly complex, and companies—especially large-scale companies—required supply of a variety of inputs, such as turbines, power stations, different types of machinery, equipment, technical expertise, infrastructure, and finances, which would potentially indicate connections to mechanical workshops as well as other suppliers. Did mining companies acquire such technical inputs from companies in Norway or did they purchase such from other countries?

---

[19] M. Badia-Miró and C. Ducoing, "The long run development of Chile and the Natural Resources curse. Linkages, policy and growth, 1850–1950", *UB Economics working papers*, 2014/318 (2014), pp. 12–13; S. Villalobos et al. *Historia de la ingenieria en Chile* (Santiago, 1990), pp. 228–229.

[20] M. N. Larsen et al., "Does FDI Create Linkages in Mining? The Case of Gold Mining in Ghana" in *Transnational Corporations and Development Policy*, E. Rugraff et al. (Basingstoke, 2009), pp. 248–249.

[21] UNCTAD, "World Investment Report 2001 Promoting Linkages" (New York and Geneva, 2001), p. 138.

[22] J. Katz et al. "Instituciones y Tecnología en el Desarrollo Evolutivo de la Industria Minera Chilena", CEPAL serie Reformas Económicas 53, 2000, p. 48.

Linkages to Mining Companies

The capital goods sector in Norway shows a different story than the ones in Latin American and African countries. Imports of machinery were reduced from the 1860s, and mechanical workshops, which grew in Norway from the 1840s, developed into a large capable and competitive industry. Bruland writes that

> [a]s the Norwegian industrial sector continued to grow in this period [… it] indicates that Norwegian producers increasingly met the needs; thus the Norwegian mechanical workshop industry succeeded in the competition with foreign producers.[23]

Production of equipment, tools, technical devices, engines, turbines, and so on increased from the mid-nineteenth century and workshops began constructing complex machinery and devices which had been previously purchased from other countries. An example illustrating this change is Nyland Workshop, which in 1892 decided to construct a machine they needed instead of importing it from England. It is found that "Norway during the nineteenth century had developed a mechanical workshop industry […whose] standard ultimately was on par with the rest of the industrialised world".[24] Actually, in the late nineteenth century the workshop industry grew faster than all the other industries in Norway. In the 1880s the industry became "a leading sector in the country".[25]

Pioneer mechanical workshops in Norway were Akers Mechanical Workshop, O. Jacobsen Machine Workshop, Kværner Works, Christiania Nail Works, Thune Mechanical Workshop, Nyland Mechanical Workshop, Trondhjem Mechanical Workshop, Bergen Mechanical Workshop, Kristiansand Mechanical Workshop, Stavanger Foundry & Dok, Trolla Works, and Mesna Works.[26] During the nineteenth century,

[23] K. Bruland, "Norsk mekanisk verkstedindustri og teknologioverføring 1840–1900", in *Teknologi i virksomhet: verkstedindustri i Norge etter 1840*, E. Lange (ed.) (Oslo, 1989), p. 41.

[24] K. Bruland, "Norsk mekanisk verkstedindustri og teknologioverføring 1840–1900", in *Teknologi i virksomhet: verkstedindustri i Norge etter 1840*, E. Lange (ed.) (Oslo, 1989), p. 72.

[25] E. Lange (ed.), *Teknologi i virksomhet: verkstedindustri i Norge etter 1840* (Oslo, 1989), pp. 17–22.

[26] F. Hodne and O. H. Grytten, *Norsk økonomi i det nittende århundre* (Bergen, 2000), pp. 198–199.

mining companies, both domestic and multinational firms, increased their use of workshops for technical services, materials, equipment, and machinery. The use of inputs by Kongsberg Silver Works illustrates the switch from foreign suppliers of technical inputs and know-how to local ones. In the early nineteenth century, Kongsberg bought machinery and equipment, such as steam engines and water column machines, mostly from companies in Sweden, England, and Germany, but from the 1830s, explosives and machine parts were purchased from domestic companies. Drammen Iron Works and Mechanical Workshop supplied Kongsberg Silver Works with five water column machines from 1872 to 1887. Kværner Works, Bærum Iron Works, and other workshops also delivered equipment. The expansion of the mechanical workshop industry correlated with this increased use.[27]

Mechanical workshops became increasingly specialised and adapted to mining companies' technical use. From the 1890s, when large-scale production, electric power, and complex mining equipment were used extensively among mining companies, workshops supplied more multifaceted tools and mechanical and electric equipment. Myren Mechanical Workshops, for instance, started to make turbines and planning machines and Kværner Workshop specialised in water power turbines.[28] The adjustments that were made by the mechanical workshop industry facilitated stronger linkages to mining companies. Workshops, in fact, delivered countless machines, equipment, and technical services to the mining sector. In the 1890s, Kværner made turbines for Røros Copper Works.[29] In 1902, the mining company Løkken Works ordered an electric system from Myren and a pump from Siemens and Haske. Kongsberg Silver Works ordered equipment and technical parts from Kværner Works, Norwegian Electric Ltd., and Borchgrevink & Company, among others, for the new electric power plant that was installed in 1914.[30] Ørens Mechanical Workshop in Trondheim provided the new flotation plant to

[27] B. I. Berg, *Gruveteknikk ved Kongsberg Sølvverk 1623–1914* (Trondheim, 1998), pp. 321–355.

[28] F. Hodne and O. H. Grytten, *Norsk økonomi i det nittende århundre* (Bergen, 2000), p. 201.

[29] F. Sæland, *Bergingeniør Emil Knudsens erindringer Inntrykk fra et bergmannsliv 1856–1897* (Kongsberg, 2005), pp. 117–118.

[30] B. I. Berg, *Gruveteknikk ved Kongsberg Sølvverk 1623–1914* (Trondheim, 1998), pp. 421 and 450.

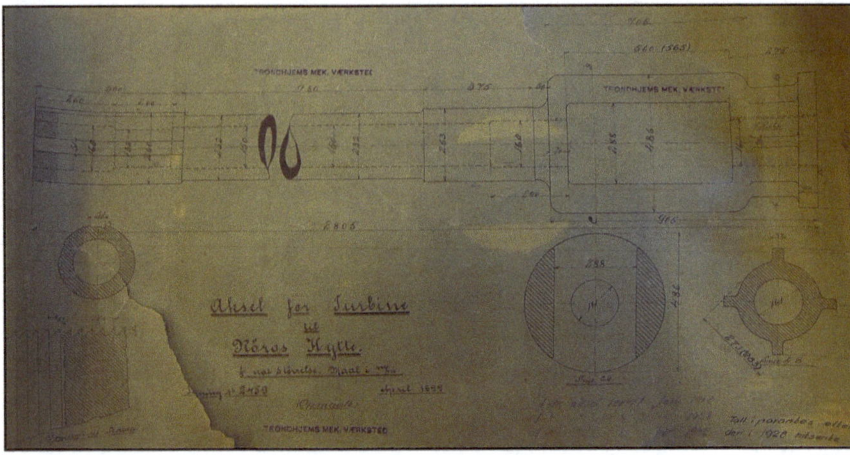

Fig. 6.1 Turbine for Røros Copper Works. Source: Trondhjem Mekaniske Verksted, Privatarkiv 310, NTNU Universitetsbiblioteket (Dora), Industritegninger, Røros Kobberverk år 1885–1909

Røros Copper Works in 1926 and the engineering firm Fer.dP. Egeberg and engineer Kraft Johansen supplied parts for another flotation plant that was used from 1932.[31]

Archival records of Trondhjem Mechanical Workshop include model drawings of numerous mining machines and equipment which were assigned to mining companies. Such drawings were the first step in the process of making technical devices, furnaces, and equipment. A wide range of illustrations of machinery, devices for electric plants, and furnaces exemplify the interactions and strong connections between mining and mechanical workshops. Illustrations were made for Røros Copper Works of turbines, a washing drum, mining trolley, converters, dynamos, a water jacket, a refining furnace, ore dressing devices, a twin blower machine, and cableways (see Fig. 6.1).[32] Gunnar Brun Nissen confirms that a double-acting cylinder blowing machine with turbine and pipeline, three converters with carriages, and other equipment were purchased

[31] G. B. Nissen, *Røros Kobberverk 1644–1974* (Trondheim, 1976), p. 247; M. Mortenson, Utviklingslinjer i norsk oppredningsteknikk, *Særtrykk av Tiddskrift for kjemi, bergvesen og metallurgi* 2 (1949), p. 6.

[32] Trondhjem Mekaniske Verksted, privatarkiv 310, NTNU Universitetsbiblioteket (Dora), industritegninger, Røros Kobberverk år 1885–1909 (three boxes).

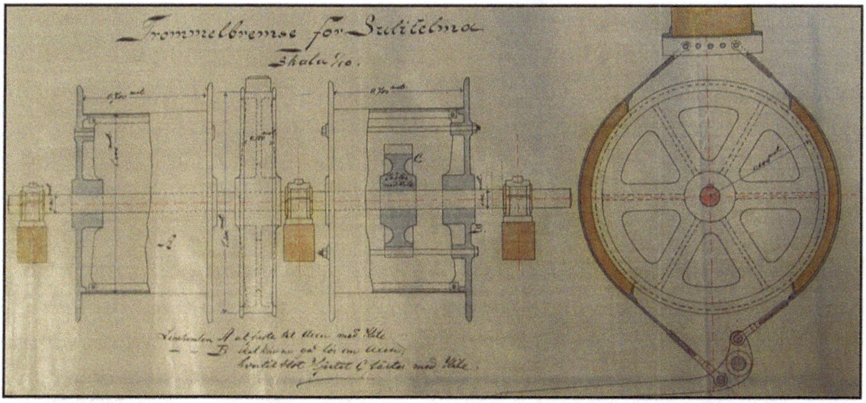

Fig. 6.2 Drum break for Sulitjelma. Source: Trondhjem Mekaniske Verksted, Privatarkiv 310 [private archive], NTNU University Library (Dora), Industritegninger, Sulitjelma Gruver 1900–1956

by Røros from Trondhjem Mechanical Workshop, which were to be used in connection with the French Pierre Manhés converting technique that was installed in 1886.[33]

Trondhjem Mechanical Workshops also made equipment for multinational mining companies. Evaporation devices and water jackets, for example, were delivered in 1924, 1936, and 1938 to the Swedish Orkla Mining Company. Moreover, the Workshop's archival records include model drawings of multiple constructions, tanks, converters, furnaces, and other devices for this company. Drawings were also made of lifts and other equipment for Fosdalens Mining Works and steel constructions were sold to this company in 1928–1929.[34] A number of drawings were made of granulators for the Canadian firm Dunderland Iron Ore Company, mechanical devices for Birtavarre Mines, and lifting devices and converters and drums for the Swedish company Sulitjelma Mining Company (see Fig. 6.2).[35]

[33] G. B. Nissen, *Røros Kobberverk 1644–1974* (Trondheim, 1976), pp. 191–92.

[34] O. Schmidt, *Aktieselskapet Trondhjems mekaniske verksted 1843–1943* (Trondheim, 1945), pp. 174–181.

[35] *Trondhjem Mekaniske Verksted*, privatarkiv 310 [private archive], NTNU University Library (Dora), industritegninger, Røros Kobberverk år 1885–1909, industritegninger Fosdalens Bergverk 1895–1936 and industritegninger Sulitjelma Gruver 1900–1956.

Increased use of local workshops did not prevent mining companies from importing equipment. In the cash books of the English mining company Bede Metal & Chemical Co., established in 1895, equipment sent from other countries was registered. In November and December 1919, for instance, Bede paid customs to Røros Custom Office for a "diesel motor".[36] The same year payments were made to Bachke & C for "forward machine from England in February".[37] Subsidiaries of multinational mechanical workshops in Norway were also used, notably the German company Siemens-Schuckert in Christiania. Bede bought a dynamite-exploding electrical machine from this company in August 1906 and an electric motor in November 1906.[38] The Swiss company Brown Boveri Ltd. was paid to repair an electric motor in October 1915 and an electric motor was bought by Siemens-Schuckert in April 1913.[39]

Yet, a systematic review of the Bede Metal & Chemical Co.'s cash books really shows the extent to which local mechanical workshops—both small and big ones—were used (see Appendix 6 of Bede's cash books). Equipment was imported sporadically, but tools, such as hammers, mine lamps, fuses, mine ladders, dynamite, chains, belts, wires, and screws, were purchased every year from local shops and workshops. More complex machinery was also acquired from local workshops: a pump and "pump accessories" were bought from W. Fischer & Sön in July 1899[40]; an "electric plant: excavation turbine station" was purchased from Jörgen Nyhus and A. A. Reitan in June/July 1905[41]; an electric installation was bought from E. Fjeldseth in September 1910; and a drilling machine was

[36] Bede Metal Killingdal, privatarkiv 107 [private archive], Statsarkivet i Trondheim, 12 kassabok nov. 1919-des. 1919.

[37] Bede Metal Killingdal, privatarkiv 107 [private archive], Statsarkivet i Trondheim, 11 kassabok nov. 1906-okt.1919; 12 kassabok nov. 1919-des. 1919.

[38] Bede Metal Killingdal, privatarkiv 107 [private archive], Statsarkivet i Trondheim, 10 kassabok sept. 1895; okt. 1906; 11 kassabok nov. 1906-okt.1919.

[39] Bede Metal Killingdal, privatarkiv 107 [private archive], Statsarkivet i Trondheim, 11 kassabok nov. 1906-okt.1919.

[40] Bede Metal Killingdal, privatarkiv 107 [private archive], Statsarkivet i Trondheim, 10 kassabok sept. 1895.okt. 1906.

[41] Bede Metal Killingdal, privatarkiv 107 [private archive], Statsarkivet i Trondheim, 10 kassabok sept. 1895.okt. 1906.

bought from Sverre Mohn in September 1918.[42] Local workshops also supplied technical services and repaired machinery. Ørens Mechanical Workshop in Trondheim, for instance, repaired one of Bede's pumps in June 1898.[43] The high usage of local workshops continued until the company ceased operation in 1946.[44]

More widely, mechanical workshops started early to make equipment, turbines, engines, machinery, and so on for a wide range of industries in Norway, including other important natural resource industries, such as timber and fish industries.[45] Historical analyses show that natural resource industries developed strong linkages to a wide range of industries, such as engineering companies, research centres, as well as the capital goods sector.[46] And linkages between natural resource industries and the strong capital goods industry continued. From the 1970s, the shipbuilding industry diversified and expanded their activities to supply equipment for the growing oil and gas sector. Other industries; engineering, information technology services, and business services, also developed to become suppliers to this sector.[47] Extensive linkages between industries in Norway continue today. This is related to the sustained growth of the capital goods industry after the turn of the twentieth century. "Industry" alone stood for 21.4 per cent of net domestic product in 1910, 22.1 per cent in 1930, and 22.7 per cent in 1935, according to official statistics. It is unclear what percentage the capital goods industry represented during these years in terms of total GDP, but it is confirmed that it was the industry's largest branch. And production of machinery and equipment is still growing. According to World Development

[42] Bede Metal Killingdal, privatarkiv 107 [private archive], Statsarkivet i Trondheim, 11 kassabok nov. 1906-okt.1919.

[43] Bede Metal Killingdal, privatarkiv 107 [private archive], Statsarkivet i Trondheim, 10 kassabok sept. 1895.okt. 1906.

[44] Bede Metal Killingdal, privatarkiv 107, Statsarkivet i Trondheim, 11 kassabok nov. 1906-okt.1919; 12 kassabok nov. 1919-des. 1919; kassabok jan. 1920-des. 1925; kassabok jan. 1926-des. 1931; kassabok jan. 1932-jul. 1938; kassabok aug. 1930-aug. 1945.

[45] F. Hodne and O. H. Grytten, *Norsk økonomi i det nittende århundre* (Bergen, 2000), pp. 198–199.

[46] S. Ville and O. Wicken, "The Dynamics of Resource-Based Economic Development: Evidence from Australia and Norway", *Industrial and Corporate Change*, Volume 22, Issue 5, (1 October 2013), p. 37.

[47] J. Fagerberg et al. "The evolution of Norway's national innovation system", *Science and Public Policy*, vol. 36, No. 6, July (2009), pp. 434–435.

Indicators, 'machinery and transport equipment' share of value added in manufacturing stood at 19.4 per cent in 1990 and increased to 29.9 per cent in 2008.[48] The growing and adaptable workshops, engineering firms, and firms specialising in technical services in Norway have facilitated key learning processes which have supported technological dynamism and innovation in mining, but also contributed to development in other natural resource industries.

Summary

The capital goods industry in Norway developed from the 1840s and became a leading branch by the end of the nineteenth century. By this time, mechanical workshops had become increasingly specialised and produced designed equipment, tools, and machinery and supplied technical services for multiple industries, including natural resource industries. They seemed to adapt heavily to companies' and industries' needs. This adaptation seemed to contribute to a shift, which consisted of mining companies increasingly using local providers for inputs instead of foreign ones. Strong linkages were created between mining companies—both local and multinational companies—and the capital goods industry. Workshops designed and produced power stations, drills, mining tools, technical services, lifts, dynamos, materials, and other smaller devices, but also offered repair services. From the turn of the twentieth century, workshops supplied more complex machinery. There are examples of mining companies importing equipment from abroad, but records indicate that most of the inputs were purchased locally. Such important linkages enabled the adoption of new mining technology in Norway in the late nineteenth and early twentieth centuries, but they have also contributed to key learning processes which have led to new specialised productions, and which have had positive knowledge effects far beyond the mining sector.

[48] F. Sejersted, *Demokratisk kapitalisme* (Oslo, 1993), p. 94; E. Lange (ed.), *Teknologi i virksomhet: verkstedindustri i Norge etter 1840* (Oslo, 1989), pp. 17–20; Norges Offisielle Statistikk, *Historisk statistikk* (Oslo, 1978), p. 96; World Data Bank, World Development Indicators (WDI) & Global Development Finance (GDF).

Bibliography

Manuscript Sources

Universitetsbiblioteket (Dora) [University Library in Trondheim, Dora]

Trondhjem Mekaniske Verksted, Privatarkiv 310 [Trondhjem Mechanical Workshop].

Johan Herman Lie Vogt, Privatarkiv Tek. 4.

Statsarkivet i Trondheim [Trondheim State Archive]

Bede Metal & Chemical Co. Ltd. Killingdal gruve, privatarkiv 107.

Secondary Sources (Literature)

Arias, M., et al. (2012). Large Mining Enterprises. Documentos de Trabajo en Economía, UCN.

Badia-Miró, M., & Ducoing, C. (2014). The Long Run Development of Chile and the Natural Resources Curse. Linkages, Policy and Growth, 1850–1950. UB Economics Working Papers, 2014/318.

Baros Mansilla, M. C. (2006). *Una historia de pioneros: Potrerillos y el Salvador.* Santiago: Quebecor World.

Berg, B. I. (1998). *Gruveteknikk ved Kongsberg Sølvverk 1623–1914.* Trondheim: Senter for teknologi og samfunn, NTNU.

Bergingeniørforening. (1949). *Tidsskrift for kemi og bergvæsen.* Oslo: Den norske ingeniørforeigning og Den polytekniske forening.

Encina, F. A. (1912). *Nuestra inferioridad Económica Sus Causas, sus Consecuencias.* Santiago: Imprenta Universitaria.

Fagerberg, J., et al. (2009, July). The Evolution of Norway's National Innovation System. *Science and Public Policy, 36*(6), 431.

Hiliart, L. (1964). *Braden Historia de una mina.* Chile: Editorial Andes.

Hodne, F., & Grytten, O. H. (2000). *Norsk økonomi i det nittende århundre.* Bergen: Fagbokforlaget.

Jones, G. (2005). *Multinationals and Global Capitalism.* Oxford: Oxford University Press.

Katz, J., Cáceres, J., & Cárdenas, K. (2000). Instituciones y Tecnología en el Desarrollo Evolutivo de la Industria Minera Chilena. *Serie Reformas Económicas 53*. Santiago: CEPAL ONU.
Lange, E. (Ed.). (1989). *Teknologi i virksomhet: verkstedindustri i Norge etter 1840*. Oslo: Ad notam forlag.
Macchiavello Varas, S. (2010). *El problema de la industria del cobre en Chile y sus proyecciones económicas y sociales*. Santiago: Imprenta Fisca de la Penitenciaria.
Mamalakis, M. J. (1976). *The Growth and Structure of the Chilean Economy*. London/New Haven: Yale University Press.
Nelson, B., & Behar, A. (2008). *Natural Resources, Growth and Spatially-Based Development: A View of the Literature*. Washington, DC: World Bank.
Nissen, G. B. (1976). *Røros Kobberverk 1644–1974*. Norway: Aktietrykkeriet.
Norges Offisielle Statistikk. (1978). *Historisk statistikk*. Oslo: Statistisk Sentralbyrå.
Pinto Santa Cruz, A. (1959). *Chile, un caso de desarrollo frustrado*. Santiago: Editorial Universitaria.
Pinto Vallejos, J., & Ortega Martínez, L. (2004). *Expansión minera y desarrollo industrial: un caso de crecimiento asociado (Chile 1950–1914)*. Santiago: Universidad de Santiago de Chile.
Ranestad, K. (2015). *The Mining Sectors in Chile and Norway*. PhD Thesis, Geneva.
Rosenberg, N. (1977). *Perspectives on Technology*. London: Cambridge University Press.
Rugraff, E., & Hansen, M. W. (2011). *Multinational Corporations and Local Firms in Emerging Economies*. Amsterdam: Amsterdam University Press.
Rugraff, E., et al. (2009). *Transnational Corporations and Development Policy: Critical Perspectives*. London: Palgrave Macmillan.
Sæland, F. (2005). *Bergingeniør Emil Knudsens erindringer Inntrykk fra et bergmannsliv 1856–1897*. Kongsberg: Norsk bergverksmuseum.
Schmidt, O. (1945). *Aktieselskapet Trondhjems mekaniske verksted 1843–1943*. Trondheim.
Sejersted, F. (1993). *Demokratisk kapitalisme*. Oslo: Universitetsforlaget.
Soto Cardenas, A. (1998). *Influencia británica en el salitre: origen, naturaleza y decadencia*. Santiago: Edit. Universidad de Santiago de Chile.
UNCTAD. (2001). *World Investment Report 2001 Promoting Linkages*. New York/Geneva: United Nations.
Villalobos, S., et al. (1990). *Historia de la ingenieria en Chile*. Santiago: Editorial Universitaria.

Ville, S., & Wicken, O. (2013, October 1). The Dynamics of Resource-Based Economic Development: Evidence from Australia and Norway. *Industrial and Corporate Change, 22*(5), 1341.

Wilkins, M. (1970). *The Emergence of Multinational Enterprise: American Business Abroad from the Colonial Era to 1914*. Cambridge: Harvard University Press.

Wilkins, M. (1974). *The Maturing of Multinational Enterprise: American Business Abroad from 1914 to 1970*. Cambridge: Harvard University Press.

Web Pages

World Data Bank. (2018, March 30). *World Development Indicators (WDI) & Global Development Finance (GDF)*. Retrieved from http://data.worldbank.org/data-catalog/world-development-indicators

7

National Geological Survey of Norway

Geological maps, prospecting, and detailed information about the content of ore, and the possibilities to make money, have been key to start up, and continue, mining projects. Geological mapping, ore surveys, and economic planning have been the basis and starting point of efficient and profitable mining. Without a deep understanding of the existing mineral and metal ore deposits, and their grade and potential profits, new mining projects can hardly take place and mining can barely advance.[1]

An increased need of mapping large geographical areas seemed to be an important reason why several states became involved in funding geological mapping and prospecting in the nineteenth and early twentieth centuries. National geological survey organisations were established with the purpose of making systematic and detailed geological maps and ore analyses of countries' resources. The utilisation of lower-grade mineral and metal ores in the nineteenth century increased the importance of detailed information about possible long-term economic returns (see Table 7.1 of national geological surveys).

In line with these initiatives, the Chilean politician and lawyer Santiago Macchiavello Varas wrote in 1923 that

[1] P. David, G. and Wright, "Increasing Returns and the Genesis of American Resource Abundance", *Oxford Journals*, vol. 6, issue 2 (1997).

K. Ranestad, *Knowledge-Based Growth in Natural Resource Intensive Economies*, Palgrave Studies in Economic History, https://doi.org/10.1007/978-3-319-96412-6_7

Table 7.1 Selected national geological surveys

Year	Country	Organisation
1835	Britain	British Geological Survey
1842	Canada	Geological Survey of Canada
1849	Austria	Federal Geological Office
1849	Spain	Commission for Geological Map of Madrid and the Realm
1851	India	Geological Survey of India
1858	Sweden	Geological Survey of Sweden
1858	Norway	National Geological Survey of Norway
1867	Italy	Geological Committee
1873	Germany	Royal Prussian State Geological Institute
1877	Finland	Geological Survey of Finland
1879	The United States	The United States Geological Survey
1888	Denmark	Geological Survey of Denmark
1896	Belgium	Geological Survey of Belgium
1910	Australia	Australian Survey Office

> [t]he utilization of the mines, which has long been done without any scientific principle, has evolved so much [...] that it can now be said that it is completely impossible to carry it out in an economically productive way, without complying with principles of physics, chemistry, geology and mineralogy.[2]

The importance of geological mapping and ore surveys for mining can be exemplified with the United States. Paul David and Gavin Wright find that between 1850 and 1950 the mining sector in this country was able to benefit from its mineral resources to a far greater extent than any other country, and they emphasise the importance of mineral prospecting in this respect. They write that "[p]rovision of geological information was perhaps the most important initial step in the collective enterprise of resource discovery and exploitation".[3] With reference to geological surveys of precious metal deposits near San Francisco, the *San Francisco Evening Bulletin* wrote in 1872 that

[2] Own translation. S. Macchiavello Varas, *El problema de la industria del cobre en Chile y sus proyecciones económicas y sociales* (Santiago, 2010), p. 26.

[3] P. David, G. and Wright, "Increasing Returns and the Genesis of American Resource Abundance", *Oxford Journals*, vol. 6, issue 2 (1997), p. 223.

> [t]hese public surveys "pay" in more sense than one, and even those who care nothing for wider and fuller knowledge for its own sake, must hereafter admit that Government expends no money more wisely and usefully.[4]

Lack of geological mapping and ore surveys has, in turn, caused stagnation of mining industries. In Chile, a lack of detailed geological maps and systematic ore surveys has been a continuous problem until recent decades, and it is used as a direct reason for the country's many unutilised ore deposits. In 1913 there were 12,403 reported inactive copper mines and the number was higher if mines with both silver and copper and gold and copper were included.[5] Two decades later, in 1931, the National Society of Mining in Chile wrote that

> [i]n all countries that have geological services and firmly established mines, the skeleton of these geological-economic studies constitutes the general geological map. Unfortunately in our country it has not been possible to carry out the preparation of this map.[6]

In 1934, it was written in the *Mining Bulletin* that the enormous non-metallic deposits in the country had not given rise to productive industries "due to the lack of scientific studies" and by the end of the 1930s, the implementation of a geological map was still "in its initial stage". This led the National Society of Mining to affirm in 1939 that the lack of scientific studies to exploit natural resources "is leading us to a dead end situation". In the same line of argument, the Credit Bank of Mining wrote that a "lack of scientific guidance in the study of these deposits causes us to make mistakes [which are] fatal for [mining] industries".[7] This problem continued for decades. An example from recent times to illustrate this is La Escondida, one of the largest ore deposits in Chile, and in the world, which was not discovered until

[4] Quote taken from T. Manning, *Government in Science The U.S. Geological Survey, 1867–1894* (Lexington, 1967), p. 11.

[5] J. Gandarillas Matta, *Estado actual de la industrial minera del cobre en el extranjero y en Chile* (Santiago, 1915), p. 75.

[6] Own translation: *Boletin de la Sociedad Nacional de Minería* (Santiago, 1931), p. 355.

[7] *Boletin de la Sociedad Nacional de Minería* (Santiago, 1939), pp. 88, 122, 300, and 310.

1981, and was then found by a private person, the North American J. David Lowell.[8]

Considering that geological mapping, prospecting, and detailed scientific ore analyses have been essential for advancing mining projects, the question here is whether the work of the NGS of Norway contributed with geological mapping and ore surveys, and whether it enabled the opening of new mining projects and the advancement of the mining sector.

Systematic Geological Mapping and Ore Surveys

Initiatives were taken from early on in Norway to increase the knowledge of the country's geological structure and mineral and metal ore deposits. Public grants, such as "scholarship for scientific trips within the Fatherland" and "Royal travel grant", as well as the "Rathke bequest"—after Professor Jens Rathke—and the "Greve Hjelmstjerne-Rosencrone bequest", were provided by the university from the early nineteenth century to mining engineers with the purpose of conducting geological excursions, mapping, and searching for ores in Norway.[9] Scholarships for such purposes continued. The geologist and mining engineer Professor Theodor Kjerulf went to Hardangervidda immediately after his studies in 1849 to carry out a geological survey with a University scholarship.[10] Ole Sandstad was given a scholarship from the Rathke Fund to study Norwegian nickel, copper, and gold deposits.[11] In 1870, Assistant Professor Hjortdal was given 50 spesidaler from the "scholarship for scientific trips within the Fatherland" to analyse the minerals in one of the tunnels of Kongsberg Silver Works.[12] After the turn of the century, in 1917, Sulitjelma Mining Company Ltd. created a fund to promote Norwegian geological knowledge. The fund paid for stays at Sulitjelma for geologists to do detailed

[8] D. De Ferranti et al. *From Natural Resources to the Knowledge Economy* (Washington, DC, 2002), p. 6.

[9] UIO, *Universitets- og skoleannaler* (Kristiania, 1834–1910).

[10] A. K. Børresen and A. Wale, *Kartleggerne* (Trondheim, 2008). 52.

[11] *Studenterne fra 1883* (Kristiania, 1908), pp. 335–338.

[12] UIO, *Universitets- og skoleannaler* (Kristiania, 1870), p. 49.

geological surveys in the Northern part of Norway.[13] These early initiatives suggest that there was an awareness when it came to developing comprehensive knowledge about geological formations and mineral and metal ore deposits in the country.

Scholarships were also provided in Norway to learn about geological structures abroad. Kjerulf went on field trips to Iceland, France, Harz and Erzgebirge in Germany, and Tirol in Austria and collaborated with researchers in Bonn and Heidelberg. In 1862, he went to Berlin, Breslau, and Vienna to inquire with specialists about geological mapping.[14] Amund Theodor Helland made several trips abroad between 1871 and 1913. In 1875, he went to Greenland and in 1876–1877 he studied microscopic petrography in Leipzig. Later he went to the lakes in northern Italy before analysing Norwegian glacial erratic movement in England, Holland, northern Germany, and Denmark. In 1879, he conducted geological surveys in the Orkney Islands, Shetland Islands, and Faeroe Islands before going to Iceland in 1881.[15] Carl Olaf Bernhard Damm obtained a Rathke scholarship and at least two other scholarships during the 1890s to study removal and transport of ore at Norwegian mining works and also to carry out geological surveys and analyses of rocks in Sweden, Germany, and Austria.[16] Such trips enabled the development of networks with geological surveys and research centres in other countries and contributed to an awareness in Norway of geological and mineralogical methods.

Public and private scholarships facilitated increased knowledge of Norway's geology and of the metal and mineral deposits that existed, but the initiative which really improved systematic geological mapping and analysis of ore was the public research centre NGS of Norway, which was established in 1858. NGS had in principle two main tasks. On the one hand, it was to contribute to new knowledge of geological features, their scope, and potential utility. On the other hand, it sought to make new and more systematic surveys of the country's geological formations and deposits. Mining engineers were heavily involved in this work.

[13] Bergingeniørforening, *Tidsskrift for bergvæsen* (Oslo, 1917), p. 71.

[14] A. K. Børresen and A. Wale, *Kartleggerne* (Trondheim, 2008). 52.

[15] J. A. Schneider, *Studenterne fra 1864* (Skien, 1916), pp. 95–99.

[16] *Studenterne fra 1886* (Kristiania, 1911), pp. 86–87.

According to the student yearbooks, 26 mining engineers worked at NGS from 1858 to 1940. Some of them conducted geological excursions and did tests on behalf of NGS during summers and others worked for longer periods as "state geologists". By the late nineteenth century, ore surveys, ore analyses, and prospecting were based on more scientific and methodologic techniques than the previously more random and unorganised methods. The adoption of new geophysical methods, surface boring tools, and later electric ore prospecting improved the work of the geologists and led to more ore findings. NGS made its work on geological maps of Norway publicly available. In 1891, a series was published, including yearbooks, academic reports, and map descriptions of Norwegian geology. Some politicians criticised NGS for working so slowly, but by 1920, geological and laboratory work by mining engineers and geologists had resulted in overview maps of the country, including Svalbard, and detailed maps of mineral deposits and ore analyses. More detailed maps were added later.[17] Other research centres which worked with, and analysed, minerals were the Public Laboratory for Raw Materials, established in 1917, which investigated the country's geological resources through chemical analysis, and the Institute for Geophysical Ore Exploration, founded in 1934, which used geophysical methods.[18]

These organised mapping and surveys contributed to discoveries of new resources and the start-up of new mining projects. Chapter 2 shows that mining productions were volatile from the late 1870s, and many of the old iron, nickel, and copper works disappeared. Production declined, much due to a recession period with low prices.[19] Thereafter, old metal projects were renewed, much due to the planning and systematic scientific research of ore deposits implemented by NGS. Børresen and Wale describe the vital role of these organisations in the start-up and development of the mining sector:

[17] A. K. Børresen and A. Wale, *Kartleggerne* (Trondheim, 2008), pp. 96 and 124.

[18] H. Carstens, ...*Bygger i Berge* (Trondheim, 2000), pp. 72–73; A. K. Børresen, *Bergtatt: Johan H. L. Vogt – professor, rådgiver og familiemann* (Trondheim, 2011); A. K. Børresen and J. T. Kobberrød (red.) *Bergingeniørutdanning i Norge gjennom 250 år* (Trondheim, 2007); AA. K. Børresen and A. Wale (red.), *Vitenskap og teknologi for samfunnet? Bergfagene som kunnskapsfelt* (Trondheim, 2005).

[19] O. Wicken, "The Norwegian Path Creating and Building Enabling Sectors", working paper, Centre for Technology, Innovation and Culture, University of Oslo, 2010, p. 15.

> The know-how the geologists at the university and NGS managed, and the rocks, ore deposits and mineral raw materials that systematically had been examined, explored and mapped, even during the difficult years for the mining industry from the 1870s, was an essential prerequisite for the market opportunities to be utilised to such an extent to quickly revive the Norwegian mining industry.[20]

Numerous new copper companies were established and the large-scale pyrite production projects were directly linked to the work by NGS.[21] The same goes for the new electro-metallurgical productions. The mining engineer Professor Johan Herman Lie Vogt notably played a big part in this. From 1889, he mapped Nordland, a region in the northern part of Norway, continuing the work which had started in the 1860s by the mining engineer T. Dahll. Vogt published his surveys and analyses of the Dunderland Valley iron ore fields, including a general description of how the minerals were formed, their geological structure, and a detailed review of the deposits, and in 1902 and 1906, the multinational companies Dunderland Iron Ore Company and Sydvaranger Mines Ltd. were founded. A similar story can be told of the marble industry. Vogt reported on several marble ores which he found represented a good foundation for a marble industry. Together with Professor Waldemar Chr. Brøgger he contributed to the start-up of Ankerske Marble Business in 1895.[22] Taking into account the importance of detailed knowledge of existing mineral deposits in a country for long-term planning and rational mining projects, without the early initiatives of making geological maps and ore analyses, and the work of NGS, the mining sector in Norway would have had a hard time advancing.

---

[20] Own translation: A. K. Børresen and A. Wale, *Kartleggerne* (Trondheim, 2008), p. 107.

[21] Some pyrite mining companies also produced copper.

[22] See A. K. Børresen and A. Wale, *Kartleggerne* (Trondheim, 2008), pp. 101–106 and A. K. Børresen, *Bergtatt: Johan H. L. Vogt – professor, rådgiver og familiemann* (Trondheim, 2011).

Summary

Without proper ore analyses and knowledge about existing mineral deposits, mining cannot take place. Evidence from Chile suggests that a lack of systematic geological surveys may have huge negative implications for the advance of mining industries and may even block the start-up of new mining projects. It is therefore important to emphasise the vital role of such work. NGS, a publicly funded organisation, established in 1858, started early with a systematic geological mapping and surveys of the mineral and metal ores in Norway. Mining engineers and geologists used advanced geological tools and chemical techniques to make extensive geological maps of the country and to implement ore analyses, which led to new knowledge of ore and the finding of new ore deposits. This extensive geological mapping of the country and prospecting of ore-rich regions led to publication series about the geological features and ore deposits in Norway. More detailed maps were made during the late nineteenth and early twentieth centuries, and they facilitated the advancement of mining projects and the establishment of new ones. Copper, silver, and nickel productions were modernised, and a new large-scale electro-metallurgical industry based on hydroelectric capacity was developed after extensive systematic geological mappings of regions with rich mineral and metal deposits.

Bibliography

Secondary Sources (Literature)

Boletin de la Sociedad Nacional de Minería. (1917). Santiago: Sociedad Nacional de Minería.

Boletin de la Sociedad Nacional de Minería. (1931). Santiago: Sociedad Nacional de Minería.

Boletin de la Sociedad Nacional de Minería. (1939). Santiago: Sociedad Nacional de Minería.

Børresen, A. K. (2011). *Bergtatt: Johan H. L. Vogt – Professor, rådgiver og familiemann.* Trondheim: Tapir akademisk forlag.

Børresen, A. K., & Kobberrød, J. T. (red.). (2007). *Bergingeniørutdanning i Norge gjennom 250 år*. Trondheim: Tapir akademisk forlag.
Børresen, A. K., & Wale, A. (red.). (2005). *Vitenskap og teknologi for samfunnet? Bergfagene som kunnskapsfelt*. Trondheim: Tapir akademisk forlag.
Børresen, A. K., & Wale, A. (2008). *Kartleggerne*. Trondheim: Tapir akademisk forlag.
Carstens, H. (2000). *...Bygger i Berge: en beretning om norsk bergverksdrift*. Trondheim: Norsk bergindustriforening Den norske bergingeniørforening Tapir.
David, P., & Wright, G. (1997). Increasing Returns and the Genesis of American Resource Abundance. *Industrial and Corporate Change, 6*(2), 203–246.
De Ferranti, D., et al. (2002). *From Natural Resources to the Knowledge Economy*. Washington, DC: The World Bank.
Gandarillas Matta, J. (1915). *Estado actual de la industrial minera del cobre en el extranjero y en Chile*. Santiago: Ed. Universo.
Macchiavello Varas, S. (2010). *El problema de la industria del cobre en Chile y sus proyecciones económicas y sociales*. Santiago: Imprenta Fisca de la Penitenciaria.
Schneider, J. A. (1916). *Studenterne fra 1864*. Skien.
Studenterne fra 1883. (1908). Kristiania.
Studenterne fra 1886. (1911). Kristiania.
UIO. (1834–1910). *Universitets- og skoleannaler*. Kristiania.
Wicken, O. (2010). *The Norwegian Path Creating and Building Enabling Sectors*. Working Paper, Centre for Technology, Innovation and Culture, University of Oslo.

Part IV

Conclusion

8

Concluding Discussion and Remarks

Countries rich in natural resources which exhibit poor economic performance are often understood as being "cursed" and recommended to shift to industries which are not based on raw materials. However, a key empirical problem with the "resource curse" argument is that some of the richest countries in the world, notably Norway, Sweden, Canada, New Zealand, and Australia, have developed fast-growing and high-performing economies based on natural resources. Similar to several African and Latin American countries, natural resource industries, notably agriculture, timber, fish, metal, and minerals have historically, and also today, constituted very large parts of these countries' economies. This shows that there are significant differences in economic performance across countries which are abundant in natural resources and which base their economies on the utilisation of raw materials. Why, then, have some natural resource–intensive economies become rich, while others have not? Scholars have recently started to address this issue in more detail and find that countries may develop *because* of their natural resources, and not in spite of them. Moreover, analyses show that natural resource industries can be knowledge-intensive and compatible with "modern knowledge-based economies" from early on. More specifically, it is found

K. Ranestad, *Knowledge-Based Growth in Natural Resource Intensive Economies*, Palgrave Studies in Economic History, https://doi.org/10.1007/978-3-319-96412-6_8

that the utilisation of natural resources has not always been based on low technological activities, but instead that natural resource industries have been highly productive, technologically advanced, knowledge-intensive, and innovative. Some natural resource industries have been based on complex and dynamic learning processes, they have created linkages to other industries within the economy and they have developed specialisations and new industries based on upstream and downstream linkages. At the same time, other natural resource–intensive economies, both large and small, have experienced lack of growth and trade difficulties.

Norway, some say against all odds, developed to become one of the richest countries in the world with a literate population from a very early stage and small social differences. By 1940, Norway's GDP per capita had reached European average and by 1950 it was higher than the European average. The extraction of oil and gas, which in recent years has constituted more than 60 per cent of exports, did not begin until the late 1960s. How did Norway develop? What are the underlying reasons for this country's high economic performance? I have sought to further our understanding of the development of this natural resource–intensive economy—which may be related to natural resource utilisation more widely—by focusing in detail on one sector, namely mining, from approximately 1860 to 1940, a period when world mining went through a technological transformation.

The argument here is that Norway's high economic performance has to a large degree been built on high-performing and dynamic natural resource industries with long historical roots. Several natural resource industries with long historical traditions became increasingly specialised and grew during the nineteenth and twentieth centuries, notably fishing, mining, timber, and timber-related industries. The mining sector, analysed here, diversified and grew considerably in the late nineteenth and early twentieth centuries. Production rose, except during downturn periods in the 1870s and 1920s, and estimations suggest that productivity of most metal and mineral productions increased dramatically from the turn of the century. The account offered here plainly only takes us part of the way in explaining why Norway, in contrast to many other natural resource–intensive economies, has experienced high economic performance. The growth in the mining sector does not alone explain the sharp economic growth that Norway experienced from the 1930s and onwards, yet its expansions and changes added to this growth. The mining sector seemed to be one of many high-performing natural resource sectors in Norway and may exemplify a more general knowledge–based development path of the country. We see, for example, that the mining sector, as well as several other industries in Norway, was largely based on foreign technology and expertise, which were absorbed and further developed by organisations in Norway. The transfer and use of foreign technology seemed to be largely due to the country's high level of education and the practical capacity to absorb knowledge. The set of organisations in Norway somehow acquired the ability to adapt to new global technological and economic circumstances.

Much research has been done on institutions and economic growth, but there is a heavy focus on general categorisations of countries which often do not reveal peculiarities and idiosyncrasies of each society, or the complexity of each country's development. Douglass North, for example, divides societies generally into "limited access societies" and "open access societies", in which he found elites controlling the economy and political institutions in the former, and free competition rules in the latter. Acemoglu and Robinson, on their part, make a simple distinction between extractive and inclusive institutions. They state that there are societies in which institutions sustain the rule of a small elite, which exploits the broader population, while other more democratic societies include the greater part of the population in development processes. The foundation of these models is institutions which either motivate innovation or constrain economic agents.

In relation to the previous, there is a general consensus that knowledge and innovation are the underlying basis of economic growth, but literature focuses less on the actual learning processes and knowledge development that might lead to innovation, and more on "prerequisites" and "incentives" to innovate. Some institutions are understood as "good", while others are "bad" for economic growth. A democratic state system, secure property rights, efficient trade systems, and widespread and high-quality education systems are some of the institutions which are seen as "good" and which are found to support learning and innovation. Authoritarian regimes, rent-seeking, and insecure property rights, in turn, are seen as "bad", and negative for economic growth. Establishing property rights, democratic systems, education systems, efficient trade systems, and so on may be important measures to encourage economic growth, but they do not, however, explain *how* innovation actually occurs. There are few analyses which actually show the direct relationship between institutions, knowledge development, innovation, and economic performance. There is a lack of evidence concerning (1) which institutions and organisations in fact have determined knowledge development and innovation, and (2) how particular institutions and organisations have enabled learning and innovation. Even innovation system approaches, which seek to examine innovation processes, can be criticised for making overly simple national models and for not capturing the complex sets of initiatives and learning processes within and across countries. The claim here has been that these general models neglect variety and the possibility that regions, sectors, and countries have grown in different ways.

How, then, can we analyse learning, knowledge development, and innovation in natural resource–intensive industries? Economic historians have in the recent decades started to analyse learning and innovation processes—and the channels through which useful knowledge for technological innovation was developed and accumulated—more directly. Institutions and organisations, such as education and technical education systems, industrial exhibitions, technical magazines and journals, research institutions, study travel by engineers and workers, foreign workers, and firms are found to be actively involved in the transfer, adoption, and diffusion of the knowledge which underpinned early industrialisation in Europe. More than providing incentives to innovate, these institutions and organisations seemed to be directly involved in actual learning and innovation processes. These economic history approaches focus attention on how knowledge accumulation occurred, and how it was transformed by learning into technological innovation within its institutional and organisational settings. This book has sought to deepen these approaches by exploring their relevance in the cases of small, natural resource–intensive economies that were far from the mainstream of industrialisation in Europe and in the United States. When we go deeper into such institutions and organisations and analyse their interplay and how they in fact have interacted with each other, how they have developed knowledge, and whether they have contributed to innovation, we can hope to understand more about economic growth.

Questions regarding the institutional and organisational settings in which the productive and growing natural resource–intensive industries in Norway have developed, the knowledge on which their development has rested, and the actors involved have been explored here through a historical empirical analysis of metal and mineral mining and extraction and the functions of key knowledge organisations in providing and spreading knowledge for this sector, namely educational establishments, mining companies, mechanical workshops, and the NGS. Student yearbooks from Norway for the years between 1855 and 1943 provide unique information about the life and work of secondary school graduates after they completed their education. This allows us to follow the graduates from school into their practices, work, and travels, and enables us to make in-depth analyses of their learning processes at school, but also their practical experiences outside school settings. Using primary sources, primarily mining engineering and technician study programmes, company records, mining journals, student yearbooks, engineering reports, and other historical written documentation, this book has sought to contribute to the literature with an analysis of the knowledge that these knowledge organisations provided, knowledge limitations, and their contributions to innovation. The state encouraged and created some of these initiatives, which means that it played a special role in the development of these organisations.

The industrialisation process that was happening in many countries during the nineteenth century created larger markets for traditional metals, as well as for new mineral products, such as steel and aluminium. In this context, mining worldwide faced major challenges in terms of finding, removing, and processing ores, in particular due to a gradual exhaustion of valuable high-grade mines and the start-up of new mineral and metal productions. Lower-grade ore in more remote areas was given attention. In order to maintain operation, and to make profits, new techniques and systems for ore prospecting, ore removal, organisation of work, and ore processing were experimented with, and many of the machinery, equipment, and techniques were developed in industrial powers such as Britain, France, Germany, and the United States. New and more powerful machinery and power sources spread around the world. Steam, animal, and manual power was gradually replaced by mechanical and electric power and enabled deeper mines and larger-scale production. New converters, furnaces, ore dressing, and smelting techniques permitted the utilisation of lower-grade ores. Norway, being a small catching-up economy, needed to develop the capability to make use of such foreign technology. Mining organisations in Norway—notably mining companies, but in collaboration with other knowledge organisations—caught up with global technological changes. By the late nineteenth century, many mining companies used mechanical power and by the early twentieth century, all companies—both domestic and multinational firms—used mechanical and electric power. Moreover, new and more efficient organisational systems and separation, conversion, and smelting techniques were used on a broad scale. These changes, in turn, permitted the diversification of the sector, the branching out of an electrometallurgical industry based on the utilisation of hydroelectric power, the introduction of large-scale production, and significant increases in production, exports, and productivity.

How can this technological transformation be explained? Based on what kind of knowledge did mining develop? How was knowledge assured, and how was learning transformed into technological innovation? Which organisations were involved in these processes? Mining was highly dependent on knowledge of the local geological ground, but in combination with in-depth knowledge of natural scientific theories and principles. Mapping, surveying, planning, drawing, and construction of underground mines involved complex calculations, measurements, and constructions, without which an in-depth comprehension of geology, mineralogy, mathematics, and mechanics would have been extremely difficult. There are multiple examples of knowledge in mineralogy, metallurgy, geology, mathematics, electro-engineering, and mechanics being used directly in working operations. One of the obvious connections to natural sciences was the extensive use of geology, chemistry, and mineralogy in the construction of geological maps and analyses of ore. Another clear example is the use of mathematical calculations and physics and mechanics in the designing and measuring of mines. The technological changes in the sector were also based on new specialisations in electro-engineering, economics, management, chemistry, and metallurgy. As mining diversified and operational processes became more complex, the knowledge bases expanded, and more specialised knowledge was used. Technologically advanced large-scale mining companies became dependent on formally trained mining engineers and technicians to manage and administrate operation and to operate complex technology. Trained workers with other educational backgrounds, such as electro-engineers, mechanical engineers, chemists, and economists, also became indispensable, as mining constructions became increasingly multifaceted and new chemical and electro-metallurgical techniques were adopted. It is emphasised here that specialised knowledge of natural sciences, engineering, and other disciplines became increasingly important to advance mining projects.

At the same time, operation was based on physical tasks of practical use of equipment, tools, and mechanical and electric devices, drawing and making of mines, removing, transporting, and processing of ores, which suggests that much of the mining activities were based on tacit knowledge. The uniqueness of each geological ground and mine, and the continuous adaptation of new techniques indicate that much of the work in mining largely involved trying and failing and learning by doing. Moreover, to be able to transfer and use machines, equipment, furnaces, techniques, and so on it was not enough to read about them in magazines or observe them at exhibitions. To understand all dimensions of technology, and especially how to select, transfer, and adopt them successfully in operation, hands-on experience on-site was key. As many of the new techniques and methods were used in other countries, learning and innovation processes in Norway involved going abroad. From such travels, engineers and technicians acquired experience with new technology, contacts, and practical knowledge of how to select, transfer, and adopt techniques. Bengt-Åke Lundvall's categorisation of knowledge into know-what, know-who, know-how, and know-why, which involve information, contacts, knowledge of how to carry out a technique, and the scientific principles for a technical problem, has been useful to identify the different aspects of knowledge in technology transfer processes in mining.

How was this complex knowledge acquired and used? Where did it come from? Knowledge was attained from many places and in a variety of ways, which makes it difficult to make clear learning and innovation processes models. However, in this case some organisations persisted and seemed to be involved more often than not. An important channel of information (know-what) was technical magazines, and the *Journal of Mining*, which was published by the Norwegian Mining Engineer Association. The function of technical journals was to spread information and publish updates about new technology and patents. International exhibitions were also important sources of information for engineers about new mining machinery, power sources, drilling, lifting, and transport equipment, converters, furnaces, turbines, and similar products on the market. Travels to exhibitions were often subsidised by public and private funds. Universities, technical schools, and research centres were the organisations which dominated the development, and diffusion, of natural scientific knowledge. Educational establishments in Norway, especially the University, the NIT, technical schools, and Kongsberg Silver Works Elementary Mining School, turned out to be key for the mining sector due to their involvement in the development of new separation, concentration, and smelting techniques in their laboratories and because they supplied engineers and technicians, and other graduates, with different types of relevant scientific and practical training. From the turn of the twentieth century, Norway seemed to have an abundance of trained workers, perhaps even too many. Without downplaying the signal effects of formal education in employment situations, this analysis has shown that knowledge that was taught in school, especially knowledge about natural sciences and engineering, was essential for certain mining procedures to be carried out. It is not clear whether formal education was a driver for technological changes in the Norwegian mining sector, but it is clear that workers with varied formal educational backgrounds supported the use, repair, and maintenance of the new and increasingly complex technology that was adopted.

Formal mining education in Norway, both at intermediate and higher levels, covered a wide range of scientific and technical subjects, notably natural sciences courses, and courses in metallurgy and mining design and construction. Reforms were made, and new courses were added in the late nineteenth and early twentieth centuries in mining machinery, construction engineering, electro-engineering, and administration and economics. The new courses that were added roughly correlated with technological changes in the mining sector, which suggests that the mining education, and the study programmes, were shaped by, and adapted to the new technology that was applied. However, the strong argument that was given by the University and the NIT was that to become fully trained and capable mining engineers, formal scientific and theoretical mining instruction should be combined with a long practice and work experience. This argument was linked to the practical nature of mining and the fact that operations depended greatly on specific knowledge of the local geology. The mining engineers and technician graduates acquired practice, in turn, by working at mining companies, either in Norway or abroad.

After graduation, most mining engineers and technicians acquired long practice. As many as 75 per cent of the Norwegian mining engineers who graduated between 1787 and 1940, and a large share of the broader set of technical and engineering graduates, studied, practised, or worked for a longer period in other countries with long mining traditions. The many travelling Norwegian mining engineers were part of a wider historical Nordic trend among workers and engineers who studied and worked, and also settled, abroad, especially in the United States. There were strong connections between organisations in Norway and mining companies, universities, and research centres in other Nordic countries, as well as Germany, Britain, France, and the United States. During these trips, mining engineers networked, acquired valuable contacts, and observed and learned about foreign technology, and this knowledge in turn was transferred and used in operational activities in Norway after their return. Travelling engineers and technicians enabled learning processes which were different than the knowledge learned in school, but not less important. The experience abroad is a key example of essential supplementary knowledge to the formal and theoretical mining instruction which was acquired outside a school setting. This practical know-how and know-who, together with the scientific and theoretical training, contributed to a local capacity to use, modify, and repair newly developed techniques and method, and sometimes led to new investments and business opportunities in Norway.

Mining companies in Norway, both domestic and multinational, facilitated practical training of engineer and technician graduates through the establishment of intern and assistant positions. The mining technicians, who acquired a more practically oriented instruction than the mining engineers, administrated mines at a few of the largest mining companies. The mining engineers, on the other hand, worked in different branches within mining. They were spread across the sector and were recruited to mining companies, as well as to academia and the NIT. They easily switched between companies specialising in iron, copper, silver, pyrite, and so on, and many of them changed jobs multiple times during their career. They typically worked at several mining companies and other organisations during their career. The mining engineers took in general three different career paths: (1) managers, middle managers, or directors at mining companies; (2) they did consultant work; or (3) they taught or did research. Some of them were involved in the start-up of mining projects. The heavy involvement of mining engineers in different parts of mining shows, on the one hand, that the work experience that they acquired was broad and diverse and, on the other hand, that the formal mining education was relevant for a number of mining organisations, and not only mining companies. These work positions generally involved supervision of workers, coordination, engineering work, design, technical decision-making, strategic planning, accounting, ore analysis, and being in charge of machinery and equipment, which in turn suggests that they were heavily involved in the transfer, management, and organisation of the new power sources, equipment, methods, and techniques that were involved in the previously described technological transformation.

Foreign investments dominated the mining sector in Norway, and multinational companies from a number of countries, such as the United States, Germany, Sweden, France, and Britain, invested in mines all over the country. Countries have sought to integrate multinational companies in different ways, but their effects on host countries have varied greatly. It is stated, however, that through regulations, education systems, infrastructure, and competitive industries, multinationals can be influenced and encouraged to integrate into the local economy, to upgrade skills, and to connect to local industries and firms. In Norway, local absorptive capacity seemed to be much higher than, for example, in Latin American and African countries, which have been used as examples here. The high absorptive capacity contributed to locally integrated multinational mining companies and the spread of knowledge through the extensive recruitment of local engineers. An important feature of the work of mining engineers is that they switched jobs between multinationals and local firms. Companies—both local firms and multinationals—together with the University, the NIT, and the NGS, developed strong networks through this widespread job-switching, and this, in turn, led to a large degree of knowledge diffusion within the sector. This knowledge diffusion, and transfer of knowledge from one organisation to another, may have contributed to the relatively equal technological level among mining companies in Norway.

Mechanical workshops, which were located in different places in the country—most of them in the largest cities—supplied new and efficient equipment, such as drills, mechanical lifts, mining tools, other materials, and smaller devices, and technical services, to the mining sector. From the end of the nineteenth century, workshops supplied more complex machinery than earlier, notably power stations and machinery. Strong linkages were created between mining companies—both local and multinational—and the capital goods industry, which enabled key learning processes, which in turn facilitated installation and repair of constantly new mining technology. By the end of the nineteenth century, the capital goods industry in Norway had become highly specialised and produced designed equipment, tools, and machinery and supplied technical services for multiple industries in Norway, both the mining sector and other natural resource industries.

Geological mapping, prospecting, and ore surveys were key for the opening and advancement of mining projects. Without a deep understanding of the existing mineral and metal deposits and their potential profits, mining and extraction could hardly take place and the mining sector could barely advance. The utilisation of gradually lower-grade mineral and metal ores from the late nineteenth century increased the importance of detailed information about possible economic returns.

Table 8.1 Simplified model of knowledge organisations involved in knowledge development in Norway

Knowledge organisation	Activities	Functions
Geological Survey of Norway (NGS): A public research centre which aimed at making geological maps of the country and ore surveys	Searched for, found, and analysed mineral and metal ores, which enabled start-up of mining projects	Assured continued mining, start-up of new mining projects, and the advance of the sector
Mining companies (domestic and multinationals): Business organisations which aimed at selling minerals and metals for profit	(1) Produced pure metals, minerals, and alloys; (2) recruited local workers, technicians, engineers, and other trained workers; (3) provided scholarships for learning purposes	(1) Ensured practical trained of mining engineers, technicians, and specialists; (2) facilitated networks and knowledge transfer within the sector by allowing for a large degree of job switching among engineers; (3) facilitated travels by engineers abroad, which were key channels to acquire contacts (know-who), and for general and practical hands-on experience of foreign technology (know-how)
Journal of Mining **(The Norwegian Mining Engineer Association):** Technical journal which aimed at spreading information and news about changes in mining and new technology	Published annual issues with news about mining from Norway and other countries to mining engineers and other workers interested in, and working in, mining	Ensured the spread of codified knowledge of new technology and other information regarding mining (know-what)

(*continued*)

Table 8.1 (continued)

Knowledge organisation	Activities	Functions
Mechanical workshops (in the country and abroad): Companies aimed at selling equipment and services to industries	Produced and delivered technical services and specialised relevant tools, equipment, engines, and other apparatus to mining companies	Contributed to the installation and repair of new mining techniques
The University and the NIT (and universities abroad): Educational establishments which aimed at developing scientific knowledge and training students on tertiary level	(1) Offered consultant work and experiments in laboratories; (2) trained mining engineers, other engineers, and other specialists for mining; and (3) provided scholarships for learning purposes	(1) Maintained close links to mining companies and facilitated the development of new techniques; (2) ensured the supply of mining engineers and other trained workers to manage, coordinate, and administrate mining companies; (3) ensured travels abroad by engineers, which were key channels to acquire contacts (know-who), and for general and practical hands-on experience of foreign technology (know-how)
Kongsberg Silver Works Elementary Mining School: Educational establishment which aimed at providing mining education on an intermediate technical level	Trained mining technicians for large-scale mining companies	Supplied mining technicians to mine manager positions at large-scale mining companies

(continued)

Table 8.1 (continued)

Knowledge organisation	Activities	Functions
Technical schools in Christiania, Bergen, and Trondheim (and smaller technical schools in other places): Educational establishments which aimed at providing technical education in chemistry, mechanics, construction, and other subjects on an intermediate technical level	Trained technicians for mining companies	Supplied technicians to different types of technical and middle management positions at mining companies

Source: Based on Chaps. 4, 5, 6, and 7

Compared to other countries, Norway started early with this work. The NGS had access to advanced equipment and hired trained mining engineers to carry out mapping and to analyse metal and mineral ores in the country. The work of the NGS facilitated the opening of new mining projects and the advancement of old ones. Copper, silver, and nickel productions were modernised, and new large-scale electro-metallurgical companies developed on the basis of systematic maps and ore analyses that had been carried out in regions with rich mineral and metal deposits.

All in all, the knowledge organisations analysed here, and their functions and implications in terms of knowledge development, are summarised in Table 8.1. But before discussing their importance further, it should be stressed that other factors than the ones explored here were involved in developing and advancing the mining sector. Capital, markets, prices, and so on of course influenced the opening of new mining projects and changes in production. To explain fully the development of Norway, the empirical analysis would need to be placed in a wider domestic and global context. First of all, these knowledge organisations were only linked to one part of the production process, which was directly concerned with locating and extracting ores. Knowledge organisations should also be analysed in connection with the capital, global markets, prices, transport systems, and linked industries. Nevertheless, learning and knowledge are essential to successful natural resource–based growth, and therefore a key starting point. Second, to fully understand the origins and developments of these organisations, it would be helpful to go even further back in time, and to explore in more detail recent decades. Third, it would be essential to explore learning and innovation on a broader scale. The activities and knowledge on which, for example, agriculture, fishing, shipping, mining, manufacture, and mining have been based, and the institutional support they have depended on, varied greatly, and this remains the case even today. Different industries have rested on different businesses, learning processes, and specialisations, which calls for detailed, and extensive, analyses and comparisons.

The knowledge organisations outlined here supported the development of key knowledge aspects in a time of technological turbulence, and formed a setting that was indispensable to innovate, catch up to up-to-date mining technology, and carry out efficient and successful mining projects in Norway. The start-up and advance of mining projects in this country were based on a range of interacting and collaborating organisations, both in the country and abroad, which together accumulated knowledge and developed technological capabilities for mining purposes. The knowledge organisations—all of them—contributed with different types of knowledge, which were useful in different ways, and they all had their functions, but also their limitations. Relationships between mining companies in Norway, the University, technical schools, the Mining School, the NIT, mechanical workshops, laboratories, technical societies, the NGS—and organisations abroad—interplayed in multiple ways. Their close ties were particularly important to enable the varied and extensive learning and knowledge accumulation on which the innovative mining sector in Norway was based.

Norway seemed exceptional in developing such knowledge-generating organisations from early on. Many countries developed similar types of knowledge organisations, but they did not always generate the same amount of knowledge capacity. In natural resource–intensive economies with less growth, notably Latin American and African countries, the local set of knowledge organisations—and the general organisational and institutional settings in the countries—seemed to be much weaker and much less able to develop knowledge and local capacities. General, mining, and technical education systems developed all over the world in the eighteenth and nineteenth centuries, but they graduated much fewer students with much less practical experience with advanced and new technology; multinational companies in Latin American countries developed into isolated enclaves; capital goods industries were underdeveloped, and geological surveys were rare. It appears that the statement made by Reich is relevant in this context: "[O]nce off the technological escalator it's difficult to get back on."[1] Examples from the United States, Sweden, Germany, and Britain, in turn, suggest that more initiatives were taken in these countries to support high-performing knowledge organisations for mining. Considering the importance of such types of knowledge organisations, and the differences in performance we see between them, it seems to be highly important to take them into account when comparing differences in economic growth between natural resource–intensive economies.

The state in Norway actively supported the efficient natural resource utilisation in different ways. It contributed to, initiated, and enabled many of the knowledge organisations and their activities. First, it facilitated trade through export regulations, it ensured property rights, and it regulated businesses. State regulations that were introduced in the early twentieth century contributed to a strong national control over the natural resources and local participation in the direction and management of companies. This suggests that the state in Norway took a particularly active role. Second, the state financed the general education system, and also the University and specialised programmes in mining and engineering. General schooling was established from the early nineteenth century on the basis of earlier teaching initiatives. These initiatives contributed to a highly literate population, which seemed to be important for a general capacity building in the country. Third, the large flow of engineers networking in other countries was a result of the large amount of public, as well as private, scholarships, and funds. The state, following the argument of Acemoglu and Robinson, was "inclusive" in that it included the broader set of the population in the education and practical learning systems through laws of mandatory primary and secondary education and

[1] Quote taken from M. Cohen and D. Levinthal, "Absorptive capacity: A new perspective on learning and innovation", *Administrative Science Quarterly*, vol. 35, Issue 1 (1990), pp. 136–137.

funding. Fourth, the decision to establish and develop a public survey makes the state indirectly involved in the modernisation of old mining projects and the development of new large-scale mining productions.

What were the mechanisms that drove these knowledge organisations in Norway to develop and accumulate knowledge? How did knowledge organisations develop this way there, and not in other countries? Looking to the debate in the press and mining journal in Norway, which included industrialists, engineers, politicians, and professors, there seemed to be a general willingness to advance mining in the country. In terms of utilising natural resources found in the ground, there are many examples from other countries of large conflicts of interests developing between different industries and business groups, both historically and today, yet such issues seemed to be less present in Norway. There might have been examples of political discrepancy concerning mining issues, and of how to proceed with mining projects, but there was a general consensus, which in turn facilitated a functioning cooperation between public and private organisations.

Mining companies, the University, the NIT, the Mining School, mechanical workshops, and the NGS in Norway were similar to each other in that they were "flexible" and adapted to new ideas and technological conditions. The NGS facilitated new mining projects, the University and the NIT modified their study programmes according to changes in the sector, and companies absorbed new innovations and experimented continuously with new and more efficient ways of mining and extracting mineral and metal ores. A key point here is that these knowledge organisations were able, and managed, to transfer and use knowledge from abroad and to adapt it successfully to local conditions. The latter was related to strong ties to organisations in other countries, which were partly established through travels and work by engineers, technicians, businessmen, and workers abroad. This, in turn was related to a particularly "outward-looking" attitude and "open" institutional context, not only formal, but also informal. For example, most of the travelling engineers and technicians seemed to acquire private or public funding, but learning about foreign techniques, taking specialised courses, and acquiring contacts must have been regarded as so important that some of them went abroad at their own expense and even without scholarships. There was an active drive in Norway to take part in new technological updates, which might be found in other Nordic countries too. These societies seemed to fit what North described as "open access societies", which according to him provide the freedom to be creative and which support innovation. But examples from Latin American and African countries indicate that such extensive and strong initiatives to develop and accumulate knowledge were far from present in all countries. The less developed mining sector, mechanical workshops, mining education, and research centres may in turn be related to them being less flexible, less adaptable, and less open to new ideas, and to lower local capacities.

Bibliography

Secondary Sources (Literature)

Cohen, W. M., & Levinthal, D. (1990). Absorptive Capacity: A New Perspective on Learning and Innovation. *Administrative Science Quarterly, 35*(1), 128.

Appendices

Appendix 1

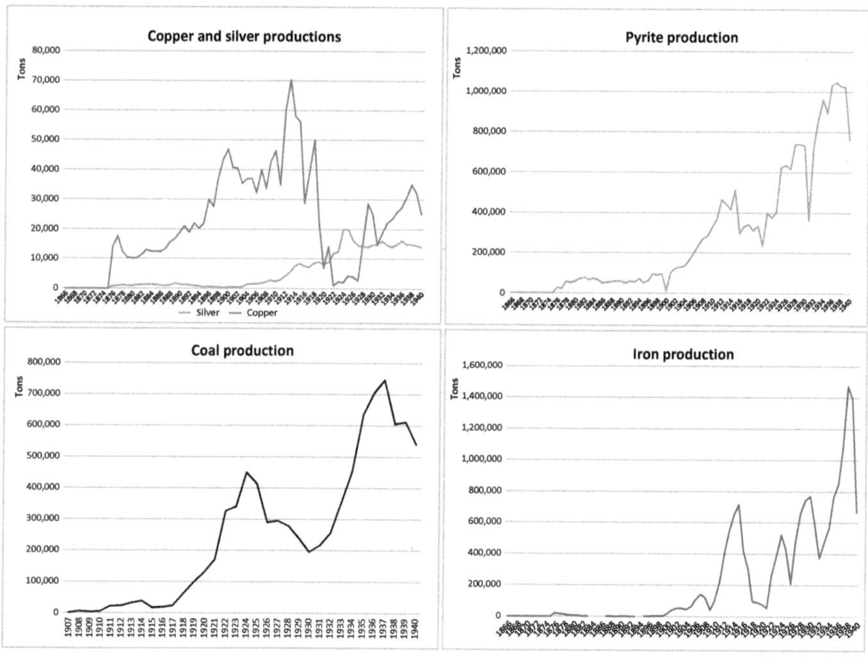

Production. Source: Norges offisielle statistikk, *Norges Bergverksdrift* (Oslo, 1866–1940)

K. Ranestad, *Knowledge-Based Growth in Natural Resource Intensive Economies*, Palgrave Studies in Economic History, https://doi.org/10.1007/978-3-319-96412-6

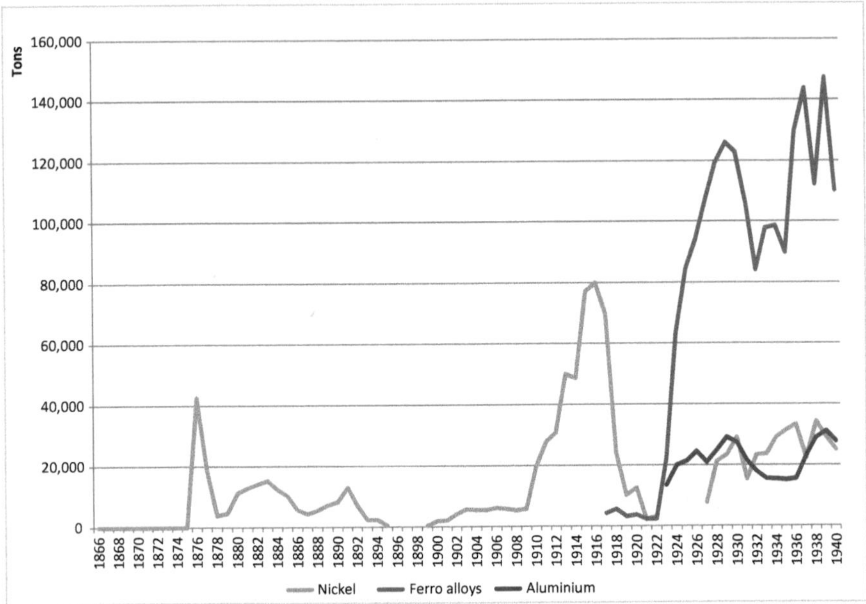

Selected electro-metallurgical products. Source: Norges offisielle statistikk, *Norges Bergverksdrift* (Oslo, 1866–1940)

Appendix 2

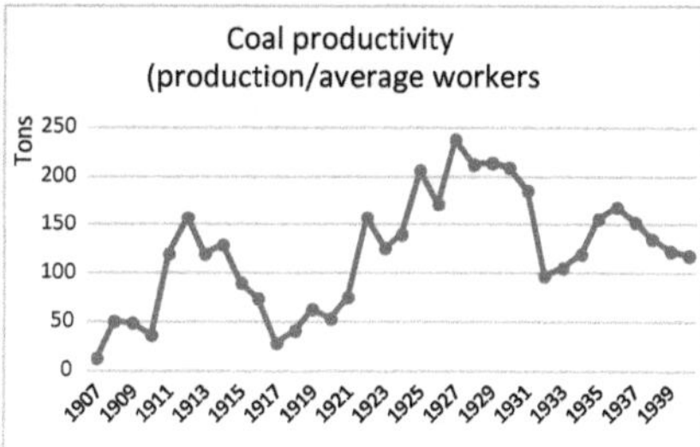

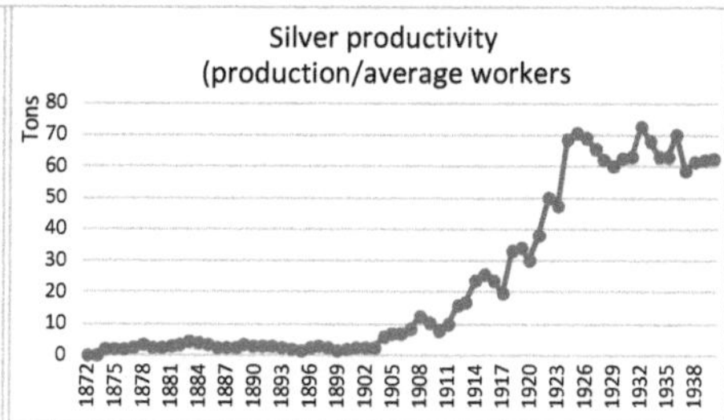

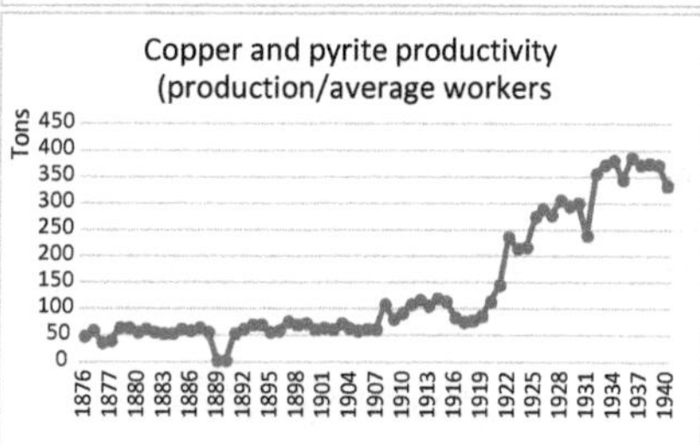

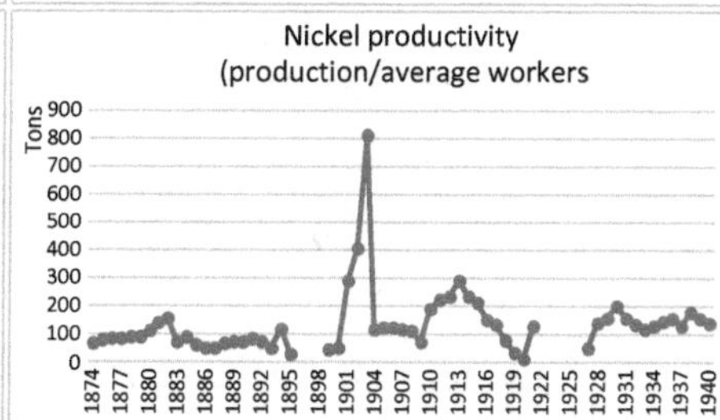

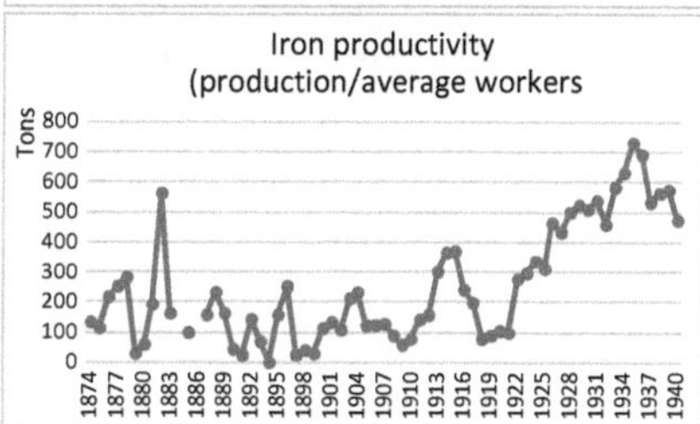

Productivity: selected mining products (metal/mineral ore). Source: Norges offisielle statistikk, *Norges Bergverksdrift* (Oslo, 1866–1940)

Average number of workers per year and total production: based on company records

Appendix 3

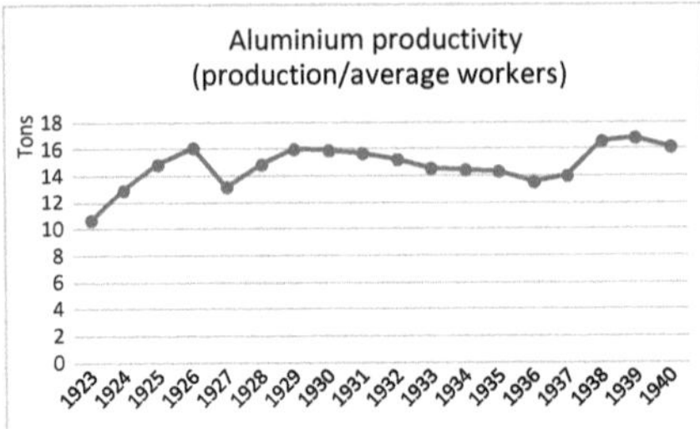

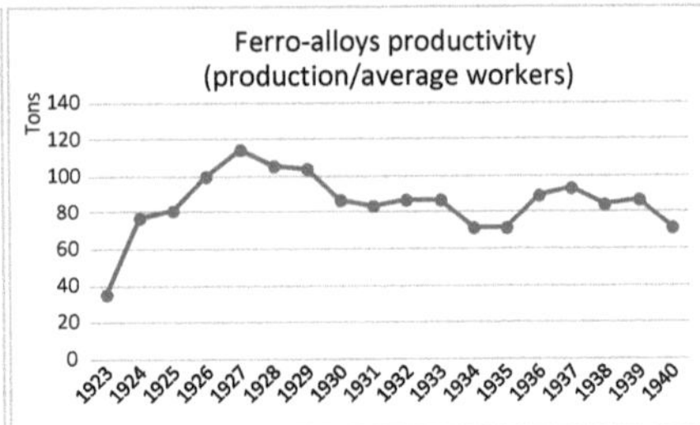

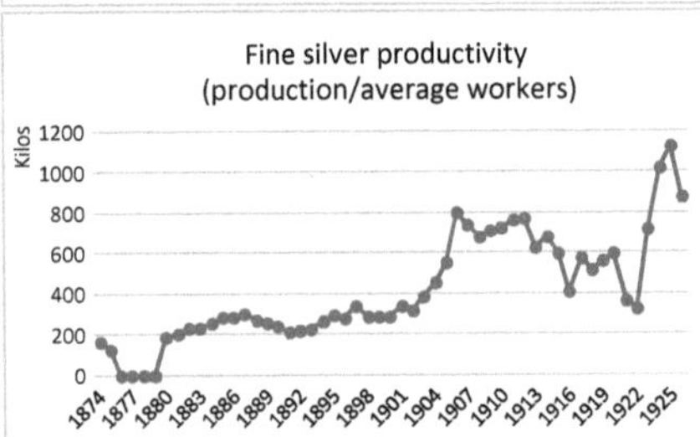

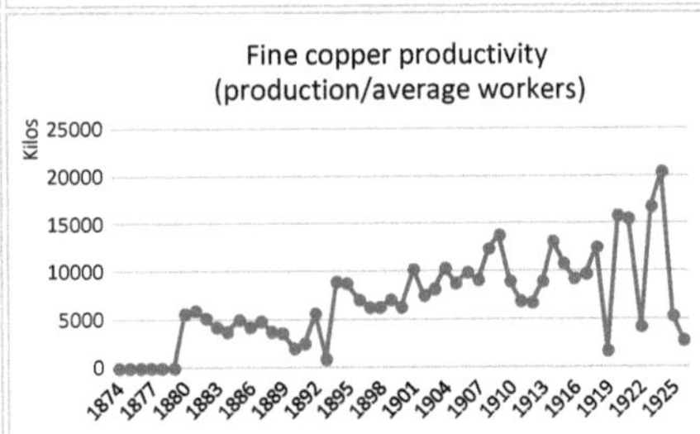

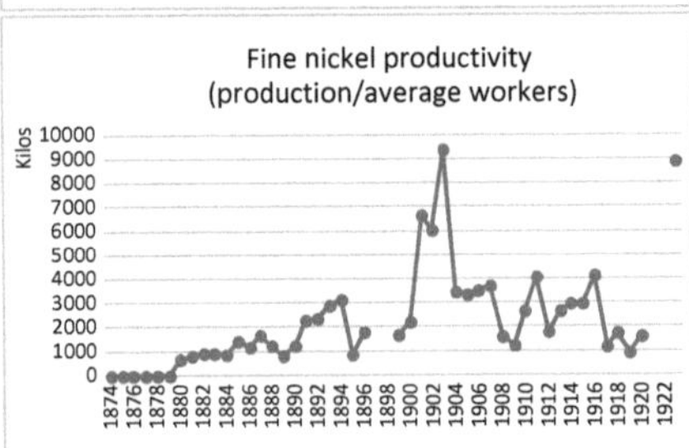

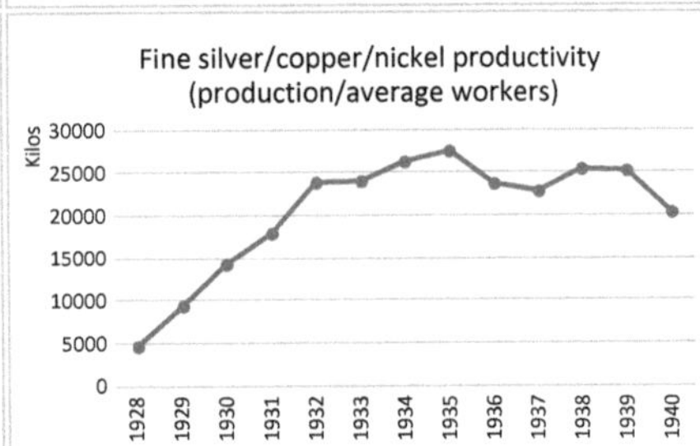

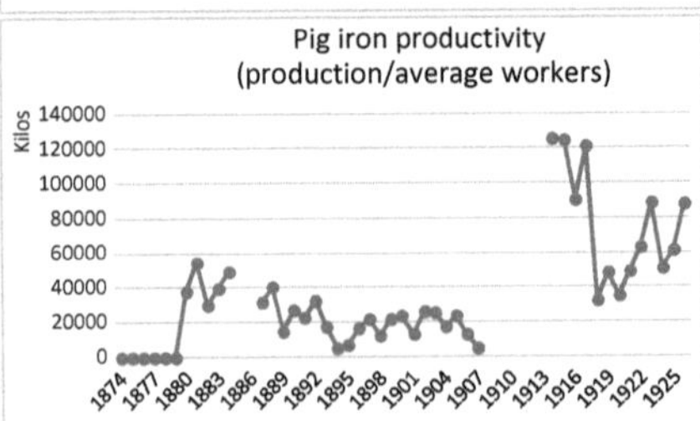

Productivity: selected metallurgical products (pure metals and alloys). Source: Norges offisielle statistikk, *Norges Bergverksdrift* (Oslo, 1866–1940)

Average number of workers per year and total production: based on company records

From 1928, workers from fine silver, copper, and nickel productions were presented together. Total production of fine silver, copper, and nickel is here divided by total workers of the three productions

Appendix 4

Mining engineers employed at mining companies in Norway, late eighteenth century to 1940
Arranged alphabetically by company

Number of mining engineers	Year employed	Company				
		Name	Number of workers[a]	Year of establishment	Nationality	Production
3	1904–1906	Alten Copper Mines	300 (max.)	1826–1878 and 1896–1909	English/ Swedish	Iron/copper
1	1892	Astrup Copper Mines				Copper
5	1850s–1918	Bamle Apatite Mines	75 (max.)		Norwegian/ French	Apatite
5	1859–1919	Bamle Nickel Works	129 (max.)	1859–1884 and 1916–1919	Norwegian	Nickel
20	Continuously	Big Norwegian Spitsbergen Coal Company	Around 590 (max.)	1916	Norwegian	Coal
4	After 1890–1914	Birtavarre Copper Works Ltd.	155 (max.)			Copper
16	Continuously	Bjørkåsen Mines	448 (max.)	1917	German	Copper/pyrite
2	1898	Blika Mines	11	1882–1901	French	Copper/silver/ pyrite
5	1820s–1898	Blue Colour Works	1300–1400 (max.)	1776 (to 1898)	Norwegian	Cobalt
1	Early nineteenth century	Bolvik (Volds) Iron Works		1692 (to 1865)	Norwegian	Iron
2	1897 and 1930s	Bossmo Mines	250 (max.)	1894–1938		Pyrite

(continued)

Number of mining engineers	Year employed	Company				
		Name	Number of workers[a]	Year of establishment	Nationality	Production
3	1927–1938	Bremanger Power Works Electric Iron Plant		1928		
1	1869	Bøilestad Copper Works (Arendal Mining and Smelting Co. Ltd)	147 (max.)	1866–1882	Foreign	Copper
2	1890s–1898	Christiania Mine Company	7 (max.)			
3	1920s	Christiania Nail Works[b]	Thousands	1853	Norwegian	Steel
3	1916–1918	Dalen Mines Ltd.				
3	1900s–1930	Dunderland Iron Mine	374 (max.)	1902	British	Iron
1	1897	Eidsvoll Gold Works	28	1758–1908		Gold
1	1912	Electro Steel Works in Jøssingfjord				Steel
6	1913–1918	Elektrokemisk Ltd.	Thousands	1904	Swedish/ Norwegian	Steel
13	Continuously	Evje Nickel Works (sold to Falconbridge Nickel Works in 1929)	350 (max.)	1872–1894 and 1908–1920	Norwegian/ English	Nickel
16	From 1829	Folldal Works	371 (max.)	1748	Norwegian	Copper
4	1905–1934	Fosdalen Mines	123 (max.)			
2	Early nineteenth century	Fritsø Iron Works	67 (max.)	1640 (to 1868)	Norwegian	Iron
1	1898	Glomsrudkollen Zink Mines				Zink
3	1862 to early twentieth century	Grimelien Copper Works	50	1862–1883	Norwegian	Copper

(*continued*)

Number of mining engineers	Year employed	Company: Name	Number of workers[a]	Year of establishment	Nationality	Production
7	Early twentieth century	Grong Mines	Around 100	1912	Norwegian	Pyrite
4	1916–1920	Gursli Molybden Mines	147 (max.)	1915–1919	Norwegian	Molybdenum
5	1898–1910	Hadeland Mining Works	77 (max.)			Zink
1	Early nineteenth century	Hakadal Iron Works	70 (max.)	1550 (to 1869)	Norwegian	Iron
1	1912	Hardanger Electrical Iron and Steel Works		1909		
1	1868	Hardanger Iron			Norwegian	Iron
1	1927	Haugsvik Smelting Works		1926	British	Aluminium
1	1862	Holtaalens Copper Works				
5	1910s–1938	Hosanger Nickel Works	86 (max.)			Nickel
1	1938	Humblen Limestone Mine	23			Limestone
1	1910s	Hundholmen Feltspat Mine				
1	1910s	Hæstad Mines				
1	1920s	Høiåsen Mines	18 (max.)	Restart 1917		Nickel
2	1927–1933	Ila and Lilleby Smelting Works		1936 (1906 and 1927)		Carbide
1	1913	Ilen Smelting Works				Carbide
1	1910s	Iron Mine in Beitstaden				Iron
2	1872–1904	Jalsberg Works	170	Before 1870		Lead/zinc
5	1910s–1939	Killingdal Mines	240 (max.)	1674	Norwegian/English	Copper/pyrite
4	1913–1932	Kings Bay Coal Mines	160 (max.)	1916	Norwegian	

(*continued*)

Number of mining engineers	Year employed	Company				
		Name	Number of workers[a]	Year of establishment	Nationality	Production
7	1903–1933	Kjøli Mine	105 (max.)	1766–1798, 1857–68, 1896–1907, 1910–1940		Copper/pyrite
1	1920s	Kjørholt Limestone Mine	77 (max.)			
75	Continuously	Kongsberg Silver Works	Around 370 (max.)	1623	Norwegian (public)	Silver
1	Around 1919	Kristian Gave Copper Works	4			Copper
16	Continuously	Kristiansand Nickel Refinery (Falconbridge Nickel Works)	577 (max.)	1910	Norwegian/ Canadian	Nickel
2	Early 1910s	Kvina Mines	27			
2	1910s–1925	Kvikne Mining Company				
1	1883	Kristiansund Mining Plant				
1	1930s	Laksådalen Molybden Mines	67 (max.)	1937		
1	1910s	Landfald Mine				
1	1910s	Langø Iron Mines	11	1609–1665 and 1882–1921		Iron
1	1867	Lindviken Pyrite Mines	23 (max.)	Before 1866		Pyrite
24	From late nineteenth century	Løkken Works (Orkla)	670 (max.)	1652	Norwegian; Swedish from 1904	Copper/pyrite
8	1906–1919	Meraker Mines	75 (max.)	1761–1920	Norwegian/ German/ Belgian	Copper/pyrite
1	1928	Mines and Ore Ltd.				

(*continued*)

Number of mining engineers	Year employed	Company				
		Name	Number of workers[a]	Year of establishment	Nationality	Production
2	1910s	Mining Company Dovre				Pyrite
1	1916	Molsaa Pyrite Mines				
1	1850	Neskilen Iorn Mines (Fossum Iron Works)	47 (max.)	End of seventeenth century to end of nineteenth century	Norwegian	Iron
1	1874	Norderhov Nickel Works	5			
1	1933	Nordic Aluminium Industry (Norwegian Aluminium Company)				Aluminium
2	1903 and 1920	Nordic Mining Company Ltd.	43 (max.)	1902	Swedish	
6	1928–1935	Northern Norway Mining Company (bergverkselskapet «Nord-Norge»)				
1	1899	Norway Mining Company Ltd. (Bergverksaktieselskapet Norge)	85 (max.)			
6	1910s–1938	Norwegian Aluminium Company (NACO)	Around 1000	1916	Norwegian/ North American	Aluminium
1	1919	Norwegian Electrical Metal Industry				
1	1917	Norwegian Electrical Steel Works		1916–1917		

(*continued*)

Number of mining engineers	Year employed	Company				
		Name	Number of workers[a]	Year of establishment	Nationality	Production
2	1920–1930s	Norwegian Metallurgical Company				
1	1909	Norwegian Mining Company Ltd. (A/S det norske bergselskap)				
1	1880	Norwegian Mining Company (Norske gruvekompanis feldstapt og apatitgruver)				Feldspar
1	1930s	Norwegian Feldspat Company Ltd.				Feldspar
1	1921	Norwegian Steel Ltd.		1823	Norwegian	Steel
2	1820s and 1937	Næs Iron Works	114 (max.)	1665	Norwegian	Iron
3	1907–1917	Ofoten Ore				
5	1924–1929	Porsa Copper Mines	72			Copper
1	1917	Porsgrund Electrometallurgical Ltd.		1913	Swiss	Steel
1	1910s	Ramsaas Nickel Works				Nickel
1	1908	Ranen Lead and Silver Works	6			
3	1909–1938	Rausand Mines	142 (max.)			
10	Continuously	Ringerike Nickel Works	261 (max.)	1849–1920	German	Nickel
1	1872	Rom Nickel Works	100	1866–1876	Norwegian	Nickel
1	1911	Rødfjellet Pyrite Mines	46	1911–1920		Pyrite
1	1918	Røen Mine				Chrome

(continued)

Number of mining engineers	Year employed	Company				
		Name	Number of workers[a]	Year of establishment	Nationality	Production
1	1916	Røragen Chrome Iron Mines				
52	Continuously	Røros Copper Works	700 (max.)	1644	Norwegian	Copper
6	1909–1920s	Røstvangen Mines	201 (max.)	1904–1921	Swedish	
1	1910	Salangen Mining Company	259 (max.)	1907	German	
1	1910s	Sargijok Gold Company				
2	1881–1885	Senjen Nickel Works	430 (max.)	1872–1886		Nickel
1	1875	Sigdal Nickel Works	39 (max.)	1874–1879	Norwegian	Nickel
3	1910s–1936	Skaland Graphite Works	45	1917	Norwegian	Graphite
1	Early twentieth century	Skandia Copper Works		1906		Copper
1	1890s	Skara Silver Mines (Kongsberg Silver Works)	60–70	1769–1782 and 1789–1810, closed 1892		Silver
1	Early twentieth century	Skorovas Pyrite Mines			Norwegian	Pyrite
1	1910s	Smorten Iron Mines				
2	1919	Stavanger Electro Steel Works		1911		Steel
11	Continuously	Stordø Pyrite Mines	367 (max.)	1907	Norwegian (public)	Pyrite
1	1907	Storolva Pyrite Mines				Pyrite
1	1908	Storvold Mines				
20	Continuously	Sulitjelma Mines	1750 (max.)	1887	Swedish/ Norwegian	Copper/pyrite

(*continued*)

Number of mining engineers	Year employed	Company				
		Name	Number of workers[a]	Year of establishment	Nationality	Production
3	1911–1919	Svanøe Pyrite Mines	Around 100	1870s–1923	Norwegian	Pyrite
2	1879–1880	Svenningdal Mines	50–60	1876–1900	Norwegian	Silver
13	Continuously	Sydvaranger Mine Ltd.	1570 (max.)	1906	German/ Norwegian	Iron
2	1910s	Tinfoss Iron Works	34 (max.)	1910		Iron
4	1918–1939	Titania Ltd.	142 (max.)	1902	Norwegian	Iron
1	1910s	Trollerud Silver Works				
1	Late nineteenth century	Tronsli Mining Company	13	1884– around 1910		Copper
9	Continuously	Vigsnes Copper Works	678 (max.)	1865	French	Copper
13	1910s–1930	Åmdal Copper Works	400 (max.)	1691	Foreign	Copper
9	1918–1940	Ørnehommen Molybdenum Mines Ltd. (Knaben Molybden Mines Ltd.)	391 (max.)	Before 1885	English	Molybdenum
8	1805–1926	Ulefos Iron Works	373 (max.)	1657	Norwegian	Iron
1	1917	Undal Works	22 (max.)	1865		
1	1917	Vadda Mines Ltd.	69 (max).	1914	Swedish	Copper
1	Early twentieth century	Varaldsø Pyrite Mines	8			Pyrite
2	1910s	Vaterfjord Molybden Mines				
1	1874	Vegaardshei Apatit Mine				Apatite
1	1910s	Vingelen and Foss Mine in Østerdalen				
4	1871–1918	Vinoren Silver Works	75 (max.)	Before 1866	Norwegian	Silver

(continued)

Number of mining engineers	Year employed	Company				
		Name	Number of workers[a]	Year of establishment	Nationality	Production
2	1877–1910s	Værdalen Nickel Works	100 (max.)			
1	1861	Ytterøen Works (the Norwegian Pyrite Company)	217 (max.)	1635–1912 (closed down several times)	Norwegian/ English	Copper/pyrite
2	1910s	Åsåren Copper Mine		Seventeenth century		Copper

[a]These are approximate numbers including functionaries

[b]Christiania Nail Works produced steel, but elaborated products (nails) which are strictly not part of mining

Sources: Norges offisielle statistikk, Norges Bergværksdrift (Kristiania/Oslo, 1866–1940); Alstad, O. (red.) Trondhjemsteknikernes Matrikel Biologiske meddelelser om samtlige faste og hospiterende elever av Trondhjems Tekniske Læreanstalt 1870–1915 (Trondhjem, 1916); Artiummatrikler studentene [student yearbooks] (Kristiania/Oslo, 1855–1940); Christiansen, H. O. et al. 25-års jubileumsberetning 1912–1937 Bergen Tekniske Skole Oslo Tekniske Skole Trondheim Tekniske Skole (Norway, 1937); Festskrift i anledning af Kristiania Tekniske Skoles 25 aars jubilæum (Kristiania, 1898); Baggethun, R. Horten Ingeniørhøgskole Horten tekniske skole En beretning om landets eldste tekniske skole gjennom 125 år (Horten, 1980); Heier, S. (red.), 100 års biografisk Jubileums-festskrift, Horten Tekniske Skole 1855–1955 (Horten, 1955); KTS, 50 årsberetning om ingeniørkullet fra Kristiania Tekniske Skole 1896 (Oslo, 1946); Kristiania tekniske skole, Ingeniørene fra KTS 1897–1947 (Oslo, 1947); Ingeniører fra Kr.a Tekniske Skole 1897 (Kristiania, 1922); K.T.S. ingeniørene av 1909: matrikkel utarbeidet til 25-års jubileet 1934: Oslo, 1934; Oslo Tekniske Skole (Oslo, 1894); KTS til Ingeniørkullet av 1910: 20 årsjubileum år 1910–1930 (Oslo, 1930); Trondhjems Tekniske Mellomskoles virksomhet i de 3 første læseaar 1912–1915 (Trondhjem, 1916); Skrift ved 50 års jubileet for ingeniørene fra K.T.S. 1894 (Oslo, 1944); L. Eskedal, BTS-matrikkelen Ingeniører uteksaminert ved Bergen Tekniske Skole 1875–1975 (Bergen 1975); TTL, 1897–1922 (Kristiania, 1922); Trondhjem tekniske læreanstalt, Festskrift ved Afslutningen av Trondhjems Tekniske Læreanstalts 25de Læseaar (Tronhjem, 1895); B. Bassøe, Ingeniørmatrikkelen Norske Sivilingeniører 1901–1955 med tillegg (Oslo, 1961); Gløersen, J. (1932): Biografiske Oplysninger om Kandidater med eksamen fra Bergseminaret på Kongsberg. Oslo: Den norske ingeniørforening; G. Brochmann (red.) Vi fra NTH de første 10 kull: 1910–1919 (Stavanger, 1934); O. Amundsen, Vi fra NTH de neste 10 kull: 1920–1929 (Oslo, 1950)

Appendix 5

Travelling Norwegian mining engineers[a] to foreign educational institutions, companies, and industrial exhibitions

Year	Name	Funding	Countries visited	Activity[b]
Around 1790	Johan Michael Kruse	Unknown scholarship	Germany Feroe Islands Scotland	Study trip to German mining works. Studied mechanics and mining construction one year at Freiberg Mining Academy
1791–1792 Around 1813	Jens Esmark		Germany Hungary Poland England	Long study trip to foreign countries; Mining Academy in Freiberg, Chemitz, mining works at Hungary, Benatet, Poland, and Schleisen Later study trip to England
1793	Christian Ancher Collett	Public scholarship	Sweden Germany	2½ years study trip to Swedish and German mining works
1794	Peter Petersen	Public scholarship	Sweden	Studying mining in Sweden, Falun, and Stockholm, during 3 years
1797	August Christian Baumann		Sweden Germany	Studied in Falun. Study trip to big German mines
Around 1800	Cajus Brandt	Fund "ad usus publicos"	Norway	Obtained 200 riksdaler annually to study coins and smelting plants at Kongsberg
Around 1800–1830	Poul Steenstrup	Public scholarship	Sweden Germany Austria	Study trip to Germany and Austria Travelled to Sweden on behalf of King Carl Johan to examine iron works
1804–1805	Peter Joachim Holten		Sweden	Study trip to Stockholm
1804–1807	Christian Münster		Germany Poland Italy	Studied at a German educational institution and travelled abroad

(*continued*)

Year	Name	Funding	Countries visited	Activity[b]
Between 1807 and 1812	Henrich Christian Strøm		Sweden France Germany England	Studied at Freiberg Mining Academy. Scientific study trips
1808	Erich Otto Knoph	On behalf of Røros Copper Works	Sweden	Travelled to Falun Copper Works with two workers from the smelting plant to study copper smelting methods
1820s 1830s	Baltazar Mathias Keilhau	Fund "scientific travels within Norway"	Norway Germany	Studied in Berlin and Freiberg. Study trip abroad for many years and many trips within Norway
1822 and 1839	Knud Olsen	On behalf of Røros Copper Works	Sweden Germany	Long trips to several German and Swedish mines
Around 1826 1846	Karl Friedrich Bøbert		Germany Belgium France Switzerland Sweden	Studied at the Mining Academy in Clausthal, University in Halle, and Mining Academy in Freiberg. Visited German mines and travelled on the Continent to study turbines
1829–1832	Amund Lammers	Public scholarship Scholarship from Kongsberg Silver Works	Sweden Germany France Austria Italy Switzerland	Studied at Falu Mining School 1830–1831. Study trip to Freiberg, Austria, Italy, Switzerland, and France
Around 1830	Knud Olsen	Scholarship from Røros Copper Works	Sweden Germany	Studied at different German and Swedish mining works and the Mining School in Falun
1837	Peter Ascanius Schult	On behalf of Røros Copper Works	Sweden	Travelled to Falun in 1837 to study wires

(continued)

Year	Name	Funding	Countries visited	Activity[b]
1837–1838	Casper Hermann Langberg		Sweden	Study trips to the Production of Coin in Stockholm
1843	Emil Bertrand Münster	Public scholarship	Germany Austria	Study trip to Germany and Austria. Studied smelting and amalgamation processes at Freiberg Mining Academy. Went also to Mansfeldt, Clausthal, and other cities to study mines and plants, chemical factories, laboratories, and collections
1843–1844	Sjur Amundsen Sexe	State scholarship	Sweden Germany	Visited mining works in Sweden and Germany and a short stay at the University in Berlin
1845	Johan Andreas Sell	On behalf of Kongsberg Silver Works	Germany	Went to Germany to study machine techniques with machine expert Professor Julius Weisbach at the Mining Academy in Freiberg. Studied turbines in Sachsen
1846	Karl Friedrich Bøbert	On behalf of Kongsberg Silver Works	Belgium France Switzerland Germany	Travelled around on the Continent to study turbines at mining works
1849 1851–1852 1862	Theodor Kjerulf	University scholarship	Norway Iceland France Germany Austria Czechoslovakia	Geological study trip at Hardangervidda, Norway. Later field trips to Iceland, France, Harz and Erzgebirde in Germany, and Tirol in Austria. Stayed with researchers Karl Georg Bischof in Bonn and Robert Wilhelm Bunsen in Heidelberg. In 1851 at the University in Bonn and other places. Study trip to Berlin, Breslau, and Vienna to inquire with specialists with regard to mapping

(continued)

Year	Name	Funding	Countries visited	Activity[b]
1846–1847 Before 1850	Tellef Dahll		Sweden Germany England	Study trip to German and Swedish mines. Went to England to learn about palaeontology and stratigraphy
1855 1877–1878	Carl Anton Bjerknes	Foreign scholarship Public scholarship	Germany France	Study trip to different places in Germany, such as the University in Göttingen. Participated at the International Exposition of Electricity in Paris in 1881. "Scientific study trip" to Germany and France
1857	Knud Hauan		Sweden Germany	Went to Falun in Sweden and Freiberg to study
1858	Hartvig Casper Christie	Public scholarship	Germany France	Studied physics at Universities in Göttingen and Paris
1858	Harald Hansteen	On behalf of Røros Copper Works	Sweden	Went to Falun
1858 1877	Carl Anthon Paaske		Germany	Study trip to Germany to study "Fahr-kunst"
1859 1876 1886 1880s	Jacob Pavel Friis	On behalf of Kongsberg Silver Works and Røros Copper Works	England Germany Sweden France Italy Spain	Went abroad to study steel drills. Went to Sweden to study the use of coke in smelting. Went with Anton Sophus Bachke to France and Italy to study electrolytic processes
1865	Poul Hansen Birch Holmsen	On behalf of Kongsberg Silver Works	Germany	Went to Germany to study water column machines for ore lifting and turbines for drainage and lifting
1867	Hans Elias Peter Andreas Holmsen	On behalf of Røros Copper Works	Sweden	Went to Swedish mines
Around 1870	Knud Hauan		Germany	Studied at Freiberg Mining Academy

(*continued*)

Year	Name	Funding	Countries visited	Activity[b]
1871	Martin Philip Rasch		Germany	Studied the use of dynamite at German mining works in Harz and Freiberg
1871–1913	Amund Theodor Helland	Public scholarship	Germany Italy England Holland Denmark	Studied mining in Switzerland and Germany 1876–1877. Study trips within Norway and abroad. Studied microscopic petrography in Leipzig, went to the lakes in northern Italy and studied Norwegian glacial erratic in England, Holland, northern Germany, and Denmark. Implemented later geological surveys in the Orkney Islands, Shetland Islands, and Faeroe Islands before going to Iceland
1873–1876	Gustav Bruun		Germany	Studied chemistry and mining at polytechnic schools in Munchen and Clausthal in Hanover
1876–1877	Abraham Godtfred Puntervold		Germany	Studied at Dresdner Polytechnikum
Around 1878	Oscar Stave		France	Studied in France
1879 1885–1889	Olaf Aabel Corneliussen	Kongsberg Silver Works Scholarship Vigsnes Works scholarship	Wales England Belgium Germany France Luxembourg	"Scientific foreign trip" in 1879 to Wales, Cornwall, Belgium, and Harzen to study mineral processing and "Fahr-kunster" at mines. In 1885 he went to the International Exposition in Antwerp in 1885 and the Universal Exposition in Paris in 1889. Studied mineral-processing plants in Germany and cable lift plants in Luxembourg
1880s	Jacob Roll		Germany	Studied in Germany

(*continued*)

Year	Name	Funding	Countries visited	Activity[b]
1880s	Anton Sophus Bachke	On behalf of Røros Copper Works	Sweden Denmark Germany Belgium England France Italy Spain	Studied pyrite and methods for cheaper processing of copper; Pierre Manhès process in France and electrolytic processes in France and Italy
1880–1900 1927–1928	Johan Herman Lie Vogt	University scholarship Unknown scholarships Foreign scholarship Two summer travel scholarships Rathke scholarship	Norway Germany Sweden England France Spain Switzerland Italy The United States	Studied at Dresdner Polytechnikum and Freiberg Mining Academy. Travelled around Europe to study mining. Went at least 14 times to Sweden. Studied iron metallurgy and quantitative-chemical analysis at the Mining School in Stockholm. Visited many Swedish mining works and institutions. Several trips to Germany and many excursions around Erzebirge and Haz. Study trip to England and Scotland to visit iron ores and coal mines. Participated at the Liverpool International Exhibition in 1886, the Universal Exposition in Paris in 1889, the Exhibition of Mining and Metallurgy in London in 1890 and the Universal Exposition in Paris in 1900
Around 1881	Diderik Cappelen		France	Study trip to France and travelled with students from Ecole des mines in Paris
1881	Truls Wiel Krefting		Germany	Finished studies at Freiberg Mining Academy

(continued)

Year	Name	Funding	Countries visited	Activity[b]
1882	Hans Knutsen		Germany Greenland	Studied at Freiberg Mining Academy. Study trip to Greenland on behalf of the King
1884–1900	Hjalmar Wilhelm Adolph Roscher	Rathke scholarship On behalf of Kongsberg Silver Works	Norway Germany France	400 kr. 2–3 months travel to Norwegian smelting plants with Rathke scholarship in 1884. Went later to study mining, Brandtske drilling machines on behalf of Kongsberg Silver Works. Went later to German and Austrian mines to study mineral processing. Participated at the Universal Exposition in Paris in 1900
1885	Christian Thaulow Nissen		Germany	Studied mining and smelting at Freiberg Mining Academy
1885–1895	Emil Knudsen	State scholarship Scholarship from Vigsnes Copper Works	The United States Sweden Germany	Studied at the Technical School in München and Freiberg Mining Academy. Assisted at the International Exposition in Antwerp in 1885 and the World Exhibition in Chicago 1893. Visited Lake Angeline Mine (iron) where he studied electric locomotive transport in the mines. Studied at the Mining Academy in Houghton and travelled to mines and ore-dressing plants in Houghton; Solway Works, Syracuse, where he studied Marwin's electrical drilling machine. He went to Sweden to visit an electric power station in Bergslagen and magnetic separation for iron ore in Stockholm
1888	Ole Sandstad	Rathke scholarship	Norway	Metallurgical studies of Norwegian nickel, copper, and coal mining works

(continued)

Year	Name	Funding	Countries visited	Activity[b]
1888 1894	Emilius Knutsen Looft		Germany	Studied in Wiesbaden and later in Leipzig
1893 1895	Emil Knudsen	On behalf of Røros Copper Works	Sweden	Went to Sweden to visit an electric power station in Bergslagen and magnetic separation for iron ore in Stockholm on behalf of Røros Copper Works Went to Germany and Schuckert's factories around 1895 in relation to the installation of the electric power station at Røros Copper Works
1894	Carl Casper Riiber	On behalf of Kongsberg Silver Works	Sweden Germany Hungary Austria	Went to study electrical drilling machines and lifts
1889	Christian Thams (construction engineer)		Norway	Participated in the Norwegian pavilion at the Universal Exposition in Paris
Around the 1890s	Henrik Kristian Borchgrevink	Rathke scholarship	Norway Canada Mexico Europe	Rathke scholarship, 200 kr. in 1891 to study removal and transport of ore at Norwegian mining works (construction methods and ore dressing). Travelled later to Canada and Mexico and around in Europe
After 1890	Holm Egeberg Holmsen	Two scholarships for technicians	Sweden Germany Austria England Scotland	Studied at Freiberg Mining Academy. Study trip to Swedish, German, and Austrian mines. Travelled later to England and Scotland. Participated at the General Art and Industrial Exposition of Stockholm in 1897

(continued)

Year	Name	Funding	Countries visited	Activity[b]
1891 Around 1900	Carl Elieson Stabell		Germany	Study trip to Freiberg smelting plants and studied smelting processes of the time. Studied later at the Technical School in Dresden
1891 1896–1897	Carl Olaf Bernhard Damm	Rathke scholarship Public scholarship Unknown scholarships	Norway Sweden Germany Austria	Rathke scholarship, 200 kr. in 1891 to study removal and transport of ore at Norwegian mining works (construction). Did surveys and studied geology, analysis of rocks, and petrography at the University in Uppsala and Heidelberg. Went also to Austria
1893	Fritz Julius Mårthén		Germany Austria	Study trips
1893 Around 1900	Andreas Holmsen	Unknown scholarships	Germany Switzerland Austria France	Obtained 200 kr. to study northern mining works. Several study trips to study geology and techniques. Studied at technical schools in Zurich and Munich and Universities in Vienna and Paris
1893–1895	John George Dahll		Germany England	Studied at Freiberg Mining Academy and Royal College of Science in London. Many study trips abroad
Before 1894	Gudbrand Thesen		Germany	Studied at Freiberg Mining Academy
1896	Richard Frederik Stalsberg	Scholarship from Røros Copper Works	Sweden	Studied at Sala in Sweden
1897	Ole Nedrum Hagen	Scholarship from Kongsberg Silver Works	Sweden	Went to the General Art and Industrial Exhibition of Stockholm and a number of Swedish mines

(*continued*)

Year	Name	Funding	Countries visited	Activity[b]
1897–1899	Ole Andreas Bachke		Germany	Studied at Freiberg Mining Academy
1897 1900	Theodor Wilhelm Holmsen (machine technician)	Two state scholarships	Norway Sweden France	Studied copper extraction at Løkken mine and went to the world exhibition in Paris in 1900 to study modern mining, especially pumps and lifting equipment. Studied ore dressing and metallurgical processes at five Swedish companies
1898	Gudbrand Henriksen		Canada	Did research on cupronickel
1899	Worm Hirsch Lund		Germany	Studied at Freiberg Mining Academy
Around 1899	Christian Horrebow Homan		England France Germany	Studied at Royal School of mines. Study trips to English, French, and German mines and smelting plants
Around 1900	Johan Carl Andresen		Europe	Several study trips
1901	Thomas Georg Münster		Sweden Denmark Germany Austria	Study trip abroad to study coin systems
1901–1906	Nils Erik Lenander		Sweden	Studied at the University of Uppsala and the Mining School in Stockholm
1901–1903 Around 1913	Lauritz Dorenfeldt Jenssen		Sweden Germany Spain The United States	Studied at Freiberg Mining Academy. Study trips
1902 1903–1904	Anton Winckler		The United States	Studied at Michigan College of Mines

(*continued*)

Year	Name	Funding	Countries visited	Activity[b]
1903–1905	Erling Lossius Jørgensen		Germany	Studied at Freiberg Mining Academy
1902–1903	Anton Martin Grønningsæter		Germany	Studied at Freiberg Mining Academy
1904–1905	Thoralf Brodtkorp		Germany	Studied at Freiberg Mining Academy
Around 1905	Sven Dahlquist		Sweden	Studied at the Mining School in Falun
Around 1905	Wilhelm Georg Tidemand		The United States	Study trips
Around 1905	Hans Henrik Horneman	University scholarship	Norway	Geological surveys in Lister and Mandal
1905 Around 1907	John Nikolai Johns		Germany Sweden	Study trip to a number of large mining companies and metallurgical plants. Studied machines in Sweden
1905–1906	Otto Fredrik Borchgrevink		Germany	Studied at Freiberg Mining Academy
1905–1906	Erik Christian Dahl Borthen		Germany	Studied at Freiberg Mining Academy
Around 1906	Carl Bugge		Scandinavia Germany	Study trip abroad to study subjects related to coins. Studied mineralogy and crystallography in München
Around 1906	Hans Helgesen		Belgium Germany Switzerland	Studied at National School of Mines in St. Etienne. Study trips
Around 1906 1911–1913 1928	Thorolf Vogt	Hjelmestjerne Rosencrone scholarship	Sweden Denmark Germany Scotland Austria	Travelled around Europe. Studied mineralogy and petrography at the University in Vienna and Göttingen. Geological expeditions to Svalbard

(*continued*)

Year	Name	Funding	Countries visited	Activity[b]
Around 1906 1934–1936	Harald Dahl		Europe Spain	Studied at Freiberg Mining Academy and later at the University in Clausthal. Travelled within Europe and visited Spanish pyrite mines
1906–1907	Arne Grønning		Germany	Studied at Freiberg Mining Academy
1906–1907	Alfred Birdy Ralson		Sweden	Studied at Falun Mining School
Around 1907–1914	Alexander Christiansen		Italy Germany Austria	Studied mining at Freiberg Mining Academy. Study trips. Secretary and organiser for the mining section at the Jubilee Exhibition in Christiania in 1914
1907–1908	Nils Christensen		Germany	Studied at Freiberg Mining Academy
1907–1908	Carl Johan Gullichsen Steenstrup	Rathke scholarship	Norway Sweden	Study trips to Norwegian, Swedish, and German mining works. Obtained 330 kr. in 1907 to study ore dressing, especially magnetic method for processing iron ore at Norwegian mining works
1907–1909	Dankert Einar Alme Torkildsen		Germany	Studied at Freiberg Mining Academy
1907–1909	Oscar Fredrik Graff		Germany	Studied at Freiberg Mining Academy
Around 1908	Eyvind Stoltz	Public scholarship Rathke scholarship	Norway ?	Many travels abroad with scholarship. 300 kr. to study the metallurgy of copper, especially modern methods for complete extraction of Norwegian copper-containing ore. In the Committee for distribution of prices at the Jubilee Exhibition in Christiania

(*continued*)

Year	Name	Funding	Countries visited	Activity[b]
Around 1908 1910s–1920s	Steinar Foslie	Public scholarship	Germany Sweden Europe The United States Canada	Study trip to Central Germany and private trips around Europe. Participated at Geology conferences in Toronto in 1913 and Madrid in 1926
1908–1910	Erik Anton Dalset	Two public scholarships	?	Studied mining abroad
Around 1908–1920	Ragnvald Støren		Norway Denmark Germany Switzerland Italy	Travelled within Norway and abroad to study chemistry, "Elmer process" (flotation and processing), ore separation, processing of pyrite, zinc, and coin system
1908	Bertel Kristoffer Skjærdal	Rathke scholarship	Sweden	300 kr. to study magnetic processing of ore
1908	Simon Smith	Rathke scholarship	Norway	Study travel to a variety of Norwegian mining works to study mining measurement and mapping of mines
1908	Simon Karenus Høegh-Omdal		Germany	Study trip
After 1908	Kristian Haslum		Norway Germany Belgium Sweden	Study travels to different mining plants in Norway, and Germany, Belgium, and Sweden. Went to Belgium and Sweden to study copper extraction
Around 1909	Harald Skappel		Germany	Studied at Freiberg Mining Academy
Around 1909 1919	Einar Hæhre		Germany The United States	Excursions with J. H. L. Vogt to German mining works and smelting plants. Study trip to the United States in 1919
1909	Hans Juell Borchgrevink		Germany	Studied at Freiberg Mining Academy

(*continued*)

Year	Name	Funding	Countries visited	Activity[b]
1909	Oscar Fredrik Graff		Germany	Studied at Freiberg Mining Academy
1909–1910	Marius Bredesen		Germany	Studied at Freiberg Mining Academy
1909–1910	Ole Andreas Rønning		Germany	Studied at Freiberg Mining Academy
1909–1911	Alf Severin Vaksdal		Germany	Studied at Freiberg Mining Academy
After 1909	Finn Holmsen Münster		Europe	Travelled to European countries to study pyrite production
Around 1910	Bjarne Hofseth		England Puerto Rico	Studied at Camborne School of Mines, England. Study trip to Puerto Rico
Around 1910–1912	Haakon Styri	The American–Scandinavian Society Scholarship	Sweden Germany Belgium France	Studied at Aachen Polytechnic University. Study trips abroad, went to Sorbonne
1910	Eyvind Flood		The United States	Studied at Denver Technical School
1910	Johan Johansson		Sweden	Studied at Falun Mining School
1910	Otto Andresen	Rathke scholarship	Norway	100 kr. to study magnetic separation of iron ore at mining works in northern Norway
1910–1911	Arne Fredrik Blom		Sweden Germany	Studied at Falun Mining School and Freiberg Mining Academy
1910 1914	Fredrik Sebastian Nannestad	Mining Fund	Norway Sweden Germany Canada	Studied modern techniques and ore processing in Europe and nickel ores in Canada

(*continued*)

Year	Name	Funding	Countries visited	Activity[b]
1910 1924–1925	Abraham Elias Kvalheim	State scholarship C. Sundt's scholarship	Germany Austria Belgium	Participated at the Universal and International Exposition in Brussels. Went to Germany and Austria to study mining. Did experiments with new methods and patents to utilise types of calibre
1911–1915	Emil Knudsen		Germany	Studied at Freiberg Mining Academy
Around 1911	Julius Helverschou	State technical scholarship	Scotland Germany Norway	Study trips to Scottish, German, and Norwegian mining districts. Studied deep drilling methods and water supply during drilling
Around 1911	Lorentz Lorch Hagen	Rathke scholarship	Norway Sweden England	Study trips abroad. 150 kr. to study metallurgy, especially electro-metallurgy at Norwegian smelting plants
Around 1911 1923	Carl Wilhelm Carstens	Public scholarship	Germany France	Studied at the University of Zurich. Travelled also to Paris
Around 1911	Rolf Havig Støre		Germany	Studied at Freiberg Mining Academy
1911	Albert Andreas Holter	Kongsberg Silver Works' travel scholarship	Sweden	Study trip to Sweden
1911 1923	Hans Ingvald Kristoffer Merckoll	Mining Fund	Sweden England Belgium Germany	Studied iron ore and processing plants in Sweden. Studied coal mines for 6 months in England, Belgium, and Germany, especially prevention of coal dust explosions
1911–1912	Kristian Refsaas		Germany	Studied electro-chemistry and metallurgy at the Technical School in Darmstadt
After 1911	Gustav Newton Kirsebom	?	Western Europe	Business travels
Around 1912	Harald Pedersen		Germany	Studied at Aachen Polytechnic University

(*continued*)

Year	Name	Funding	Countries visited	Activity[b]
Around 1912	Bjarne Reidar Saarheim		Sweden Belgium France England	Study trips
Around 1912	Sverre Blekum		Germany Sweden	Studied the molybdenum industry
1912	Sigurd Rudie		Germany	Studied at Freiberg Mining Academy
1912	Johan Falchenberg Stadheim	Rathke scholarship	Norway ?	Study trips within Norway and abroad. 100 kr. To study mining at mining works in northern Norway
1912	Peder Ytterbøe	Rathke scholarship	?	100 kr. To study electric iron smelting
1912–1913	Einar Dahl		Sweden	Studied mining and metallurgy at Filipstad Mining School.
1912–1916 1919	Nils Hofman Aall		Sweden France Germany	Studied at the Technical University in Zurich and the Technical University in Stockholm. Study trips to France and Germany
Around 1913	Georg Hauschildt Steen		Scandinavia Europe Germany Austria England	Many trips around Norway and European countries
Around 1913	Otto Falkenberg	Unknown scholarship	Scandinavia Germany Spain The United States Canada	Studied at Clausthal Mining Academy and Berlin. Study trips

(*continued*)

Year	Name	Funding	Countries visited	Activity[b]
Around 1913	Per Adam Petterson	State technical scholarship	Germany England Italy The United States North-Africa	Study trip
1913–1916	Georg Tysland	The American–Scandinavian Foundation's scholarship	France The United States Sweden England Germany	Studied electro-engineering and metallurgy at the University of Liège. Study trips abroad. Studied electric iron smelting in Pittsburgh, the United States
1913 1920	Johan Asmus Lenschow	Scholarship	Sweden Germany	Study trip to Swedish and German mines to study mining engineering and processing plants
After 1913	Albert Carinus Sunde		Germany England America China	Study trips
Around 1914	Wolmer Marlow		Belgium Holland Germany Sweden	Study trips
1914	Anders Kristian Olsen		Siberia Mongolia	Participated in Ørjan Olsen's expedition to Siberia and Mongolia
1914	Adam Frans Wilhelm Faye		Germany	Studied at the Aachen Polytechnic University
1914 1920	Fredrik Hurum		Germany The United States	Studied at the Aachen Polytechnic University and MIT Boston

(continued)

Year	Name	Funding	Countries visited	Activity[b]
1915 1928	Gunnar Aasgaard		Sweden Germany Poland Holland	Study trip to study general development in machinery and constructions as well as specific machines. Went to coal, ore mines, workshops, as well as new and old power stations in Ober-Schleisen and the Ruhr district and other places
1916–1917	Rolf Mastrander	Norsk Aluminium Co.	The United States	Went to Montana to study oil deposits and Latin America and Holland to search for ore deposits
1916 1925	David Vaage	The Int. Labour Office	Czechoslovakia Germany France Belgium England	Study trip in 1916 to Mährisch Ostrau to study electrical mining plants. During summer 1925, went to the mining districts in Harz to study mineral processing. Stayed for a while at the coal districts in France, Belgium, Germany, and England to study safety techniques in coal mines for the Int. Labour Office
Around 1916	Thorodd Wangensteen Messel		Germany Europe	Studied at Freiberg Mining Academy. Travelled around Europe
Around 1916	Harald Severin Diderik Cappelen		Germany	Studied at Freiberg Mining Academy
1916	Gunvald Birger Thorkildsen	State scholarship	Germany The United States Canada	Studied at Freiberg Mining Academy. Study trip to the United States and Canada to study modern mining
1916 1920	Leif Lyche		Germany	Studied at the Technical School Breslau

(*continued*)

Year	Name	Funding	Countries visited	Activity[b]
Around 1917	Arne Drogseth		Sweden Germany Switzerland Russia	Studied "technical subjects"
Around 1917	Carl Gottfred Larsen		England	Studied at Royal School of Mines in London
1916–1917	Sverre Winnem		Germany	Studied at Freiberg Mining Academy
1917	Truls Wiel Graff		Germany	Studied at the Aachen Polytechnic University
1917–1922	Erling Robsahm		Germany	Studied at the Technical School in Darmstadt, the Aachen Polytechnic University and Breslau
1917 1924	Trygve Brodtkorb		Sweden England The United States	Study trip to mining works
1918 1922	Appollonius Lijedahl Rosenlund	On behalf of Kristiansand Nickel Refinery	Canada Sweden	Visited nickel mines in Ontario in Canada and site investigations used Swedish state railways
Around 1918	Einar Trøften		Central Europe Sweden England	Study trips
1919	Ferdinand Peder Egeberg		Germany The United States	Studied at Freiberg Mining Academy. Study trip to the United States to study modern techniques, especially flotation
1919	Erland M. Lewin		England	Studied at Durham University
1919	Finn Andersen		England	Studied at the Camborne School of Mines in Cornwall
1919	Otto Hansen		Britain	Studied in Glasgow
1919	Erik Einar Haag		Sweden	Studied at Falun Mining School

(continued)

Year	Name	Funding	Countries visited	Activity[b]
1920	Sigurd Westberg		The United States	Studied at the University of Pittsburgh
1920	Halvard Dale	Orkla Fund	Germany Austria	Metallurgical studies
1920–1921	Brynjulf Dietrichson	Orkla Fund	France Belgium Spain	Study trip to France and Belgium. Stayed at mining works in France and Spain
1921	Werner Carl Ferdinand Poensgen		Germany	Studied at the Aachen Polytechnic University
1921	Karl Johan Hansen	Scholarship	Sweden Denmark	Studied technical schools in Sweden
Around 1921	Gunnar Schjelderup		Germany The United States	Studied electro-chemistry and metallurgy at the Technical School in Dresden. Studied practical operations at steel works
1921–1922	Enok Willmann		Germany	Studied at Freiberg Mining Academy
1921–1922	Leif Hartmann Støren		Germany	Studied at Freiberg Mining Academy
1922	Torkell Berntin Gjerstad	State technical scholarship	Germany	6-month stay at Krupp industries in Germany
1922	Johan Askeland (unknown)	Orkla Fund	Germany	Study trip to Germany to study modern mining and ore dressing
Around 1923	Arvid Thunæs		Latin America Europe Canada	Travelled around
1923	Alf Ihlen		The United States	Studied at MIT in Boston
1923	Johan Engelhart Braastad		Germany	Studied at the Aachen Polytechnic University

(continued)

Year	Name	Funding	Countries visited	Activity[b]
1923	Rolf Falck-Muus		England France Germany Denmark	Study trips
1923	Birger Egeberg	State scholarship	England France Germany	Studied at the Aachen Polytechnic University. Study trip to English iron and steel plants and visited France, Germany, and England to study silver industry
After 1923 1937–1940	Magne Mortensen	Orkla Fund Norwegian Technical Institute scholarship	Norway	Several study trips. Visited Norwegian mines to study ore geology and electrical conditions and ore dressing. Studied also flotation processes
Around 1924	Henning Mastrander	Orkla Fund Norway–America Fund	England Holland Belgium Germany The United States	Studied mining and entrepreneurship. Studied at the University of Columbia. Participated at congresses and visited modern American iron and coal mines and studied the application of new mining machines, loading facilities, and electric machinery
1924	Arne Okkenhaug	Orkla Fund	Sweden Germany	Studied mining
1924	Bjarne Askeland	Orkla Fund	Sweden	Study trip
1924	H. Chr. Jessen	Orkla Fund	The United States	Studied American ore-dressing and mining techniques
1924 and 1927–1928	Gunnar Horn	Norwegian Military Goods Insurance Fund	Germany Poland	Study trips and stay at Technissche Hochschule og Preussische Geologissche Landesanstalt, Berlin

(continued)

Year	Name	Funding	Countries visited	Activity[b]
1924 and 1931	Steinulf Smith-Meyer	State technical scholarship Orkla Fund	Germany Canada The United States	Study trip to German coal mining district. Study trip to mining works in Canada and the United States to study flotation techniques
1925–1927	Bjørn Fougner		Germany	Studied at the Technical School in Breslau
1925–1926, 1932–1933, and 1935	Einar Sverdrup		Germany England France	Several study trips
1926	Einar Slaatto	Orkla Fund	England The United States Germany Holland	Study trips in Europe and studied mining and coal processing in the United States
1926–1948	Aage Christensen		Germany The United States	Studied at Freiberg Mining Academy. Study trip to a variety of mines
1927–1930	Joakim Lindholm	Orkla Fund	Germany Belgium The United States	Study trip abroad to study coal mining, ore dressing and administration
1928–1929	Johan Kraft Johanssen	Orkla Fund	The United States	Study trip abroad to study mining and ore dressing
Around 1930	Henrik Steffens Hagerup Jenssen		Sweden England Germany The United States	Several study trips
1931	Haakon Brækken	Mining Fund	The United States	Studied the development of geophysical methods of surveying

(*continued*)

Year	Name	Funding	Countries visited	Activity[b]
1931	Robert Major Brun	Orkla Fund	Sweden Finland The United States	Studied mining and ore dressing
1931–1932	William Straube		Germany	Studied at Freiberg Mining Academy
1932 1934	Per Munthe-Kaas Sandvik		England Sweden Germany	Study trips
1933	Johan Fredrik Gørrissen		Germany	Studied at Aachen Polytechnic University
1933–1934	Per Sandved	Orkla Fund	Germany Sweden Czechoslovakia	Visited mines, smelting plants, and laboratories
1933–1934	Arne Carlsson	Orkla Fund	Sweden Germany Austria Czechoslovakia Italy	Studied for a while in Sweden and Germany
1933–1934 1937	Harald Nordrum Ross	Orkla Fund	England Holland Germany	Studied modern coal mining in Germany and England
1934	Hans Helge Schou		The United States	Studied at Michigan College of Mining and Technology
1934	Jarle Kuvås		Sweden	Studied at Goteborg Technical School
1934–1935	L. Breder (not mining engineer)	Norwegian Technical Institute scholarship	Norway	Travelled around Norway to study mining plants

(*continued*)

Year	Name	Funding	Countries visited	Activity[b]
1934–1938	Christian Hiorth Aall		France	Studied in Grenoble and travelled around in France
1934 1936–1939	Hilmar Kragh When	C. Sundt's Scholarship	Germany Austria Hungary Czechoslovakia Poland Holland Belgium France	Studied at Freiberg Mining Academy. Study trips abroad
1935	Herbert Olav Bronder	Hs fund	Sweden Denmark Germany England	Study trips
1935	Jan Herman Reimers		Germany	Studied at the Technical University in Karlsruhe
1935–1936	Arne Hofseth	Orkla Fund	The United States	Studied geophysics at the Colorado School of Mines. Studied geophysical ore survey abroad.
Around 1936	Arne Boye Holt		France The United States	Studied physical metallurgy at School of Mines in Mons, organic chemistry at University of Columbia, and organic and physical organic chemistry at Polytechnic Institute of Brooklyn
1936	Ragnar Christoffersen	Orkla Fund	Sweden Germany	Studied mining
1936	Thorstein Kavli		Germany	Studied at Aachen Polytechnic University

(*continued*)

Year	Name	Funding	Countries visited	Activity[b]
1937	Einar Falkum (chemical engineer)	Norwegian Technical Institute scholarship	Svalbard	Studied utilisation of coal in Svalbard
1937	Einar Olav Evensen	?	Germany	Studied at Aachen Polytechnic University
1937	Erling Rytterager	?	Germany	Studied at Aachen Polytechnic University
1938	Bjarne Holmsen	?	The United States Canada	3 months study trip to mines
1938–1939	Worm Lund	Orkla Fund	?	Studied iron processing
1938–1939	Thor Matheson Amdal	Orkla Fund	Sweden	Studied mechanical loading in modern mining
1940	Olav Bergersen	?	Germany	Studied at Aachen Polytechnic University
1940	Christian Fredrik Sontum	?	The United States	Studied at the University of California

[a]These trips do not include geological excursions in Norway, which were continuously implemented by professors and students at the University and the NIT

[b]In some cases (around five) information about the engineer and the funding is known, but it is unknown whether the engineer actually carried out the trip or not

Sources: Norges offisielle statistikk, *Norges Bergværksdrift* (Kristiania/Oslo, 1866–1940); Alstad, O. (red.) *Trondhjemsteknikernes Matrikel Biologiske meddelelser om samtlige faste og hospiterende elever av Trondhjems Tekniske Læreanstalt 1870–1915* (Trondhjem, 1916); Artiummatrikler studentene [student yearbooks] (Kristiania/Oslo, 1855–1940); Christiansen, H. O. et al. *25-års jubileumsberetning 1912–1937 Bergen Tekniske Skole Oslo Tekniske Skole Trondheim Tekniske Skole* (Norway, 1937); *Festskrift i anledning af Kristiania Tekniske Skoles 25 aars jubilæum* (Kristiania, 1898); Baggethun, R. *Horten Ingeniørhøgskole Horten tekniske skole En beretning om landets eldste tekniske skole gjennom 125 år* (Horten, 1980); Heier, S. (red.), *100 års biografisk Jubileums-festskrift, Horten Tekniske Skole 1855–1955* (Horten, 1955); KTS, *50 årsberetning om ingeniørkullet fra Kristiania Tekniske Skole 1896* (Oslo, 1946);

Kristiania tekniske skole, *Ingeniørene fra KTS 1897–1947* (Oslo, 1947); *Ingeniører fra Kr.a Tekniske Skole 1897* (Kristiania, 1922); *K.T.S. ingeniørene av 1909: matrikkel utarbeidet til 25-års jubileet 1934:* Oslo, 1934; *Oslo Tekniske Skole* (Oslo, 1894); *KTS til Ingeniørkullet av 1910: 20 årsjubileum år 1910–1930* (Oslo, 1930); *Trondhjems Tekniske Mellomskoles virksomhet i de 3 første læseaar 1912–1915* (Trondhjem, 1916); *Skrift ved 50 års jubileet for ingeniørene fra K.T.S. 1894* (Oslo, 1944); L. Eskedal, *BTS-matrikkelen Ingeniører uteksaminert ved Bergen Tekniske Skole 1875–1975* (Bergen 1975); TTL, 1897–1922 (Kristiania, 1922); Trondhjem tekniske læreanstalt, *Festskrift ved Afslutningen av Trondhjems Tekniske Læreanstalts 25de Læseaar* (Tronhjem, 1895); B. Bassøe, *Ingeniørmatrikkelen Norske Sivilingeniører 1901-55 med tillegg* (Oslo, 1961); Gløersen, J. (1932): *Biografiske Oplysninger om Kandidater med eksamen fra Bergseminaret på Kongsberg*. Oslo: Den norske ingeniørforening; G. Brochmann (red.) *Vi fra NTH de første 10 kull: 1910–1919* (Stavanger, 1934); O. Amundsen, *Vi fra NTH de neste 10 kull: 1920–1929* (Oslo, 1950); B. I. Berg, *Gruveteknikk ved Kongsberg Sølvverk 1623–1914* (Trondhjem, 1998); V. Bjerknes, *C.A. Bjerknes Hans liv og arbeide* (Oslo, 1925); G. A. Blom, *Fra bergseminar til teknisk høyskole* (Oslo, 1958); A. K. Børresen and A. Wale, *Kartleggerne* (Trondhjem, 2008); *Det Kongelige Norske Frederiks Universitets Aarsberetning* (Kristiania, 1840–1911); R. Falck-Muus, *Bergmannsutdannelsen i gamle dager, norske bergmenn til Sverige som ledd i utdannelsen* (Oslo, 1949); G. B. Nissen, *Røros Kobberverk 1644–1974* (Trondhjem, 1976); F. Sæland, *Bergingeniør Emil Knudsens erindringer Inntrykk fra et bergmannsliv 1856–1897* (Kongsberg, 2005); Bergingeniørforening, *Tidsskrift for bergvæsen/Tidsskrift for kemi og bergvæsen* [Mining Journal] (Oslo/Kristiania, 1913–1940); Norges Tekniske Høiskole, *Beretning om virksomheten...* (Trondhjem, 1920–1940); Den tekniske høiskole i Trondhjem *Program for studieåret...* (Trondhjem, 1911–1940); *Teknisk Ukeblad* [Technical Magazine] (Oslo, Kristiania, 1898); *Universitets- og skoleannaler* [University and school annals] (Kristiania, 1834–1916)

Appendix 6

Purchased equipment and technical services by the Bede Metal & Chemical Company (Killingdal Mines), selected years[a]

Year	Technology	Country of origin	Supplier
1895–1896	"window panes", "leather", "miners candles"	Norway	A. Haugan
	"20 m tape"	Norway	A. Motzfeldt
	"laboratory clothes", "dishes for laboratory"	Norway	C. Rönning
	"bar iron", "steel hammers", "tools"	Norway	Chr. Thaulow & Söns
	"stores and machinery"	Norway	Emil Grönning
	"construction"	Norway	Erik Schjölberg
	"various tools", "lamps and post bag"	Norway	G. A. Hartmann
	"brooms"	Norway	H. O. Grönli
	"hammer shafts"	Norway	J. Rönningen
	"hammer shafts"	Norway	Lars Josvold
	"lamp glasses"	Norway	Lars Skancke
	"hinges and screws"	Norway	M. Engzelius & Sön
	"new store buildings"	Norway	Ole Lundemo
	"materials"	Norway	P. I. Ramla
	"telephone poles"	Norway	P. Krokan
	"lamp glasses"	Norway	P. O. Rösten
	"materials"	Norway	Reitan
	"glass", "stores"	Norway	Aalens Övre Forbrugsforening
	"steam whistle (machinery)"	?	?

(*continued*)

Year	Technology	Country of origin	Supplier
1900–1901	"oils skins"	Norway	A. Kroglund
	"mine lamp"	Norway	A. Motzfeldt
	"9 boxes chalk"	Norway	A. Reitan
	"piping (steam) machinery", "crow bars", "hammers"	Norway	Chr. Thaulow & Söns
	"fuse"	Norway	Engzelius & Son
	"lamps", "50 brooms"	Norway	Emil Grönning
	"140 hammer shafts (large) 90 small"	Norway	Erik Grönli
	"lamp wick"	Norway	G. A. Hartmann
	"spikes kvs. 50", "kvs. 100 Rall spikes", "rail spikes"	Norway	Gustav Aspelin
	not specified	Norway	H. & F. Bachke
	"lamp glass"	Norway	J. J. Siem
	"12 ladders"	?	John S. Öien
	"hammer shafts"	Norway	Lars Fosvold
	"Kvs 300 dynamite", "fuse"	Norway	M. Engzelius & Sön
	"500 rings fuse"	?	Nora Fuse Manufactory
	"36 mine lamps", "30 miner's lamps"	Norway	O. A. Moxness
	"hammer shafts" large and small	Norway	Ole J. Grönli
	"stoves: renewals"	Norway	P. O. Jenssen
	"ladders", "ladders-142 rungs"	Norway	Peder Kirkhus
	"India rubber packing"	Norway	W. Fischer & Sön
	"12 empty barrels"	Norway	Öien & Wahl
	"steam irons (?) machinery", "nails", "lamps", "iron lamps", "hammers", "dynamite", "fuses", "detonators", "steel rope", "tools", "shovels", "iron pails", "hammer shafts", "nitroglycerine", "krafsers", "trugs"	?	?

(*continued*)

Year	Technology	Country of origin	Supplier
1909	"planed and grooved boards"	Norway	A. E. Gildseth
	"packing, dynamo, tubing", "driving belts, bends", "pulley", "screws, lead tubing", "brass lock"	Norway	Albert E. Olsen
	"spades", "spades and hammer handles"	Norway	Axel Nielsen
	"electric plant"	Norway	Annual government inspection
	"raw hide pinion (wheel) and copper wire", "cable", "safety fuses", "bronze", "repair of motor", "repair of 2 telephones", "telephone effects", "telephone reserve pieces"	Switzerland (Norwegian subsidiary)	Brown Boveri
	"wire rope", "200 screws", "asphalt roof paper", "bolts", "asphalt paper", "roof paper", "nails", "rivets", "iron", "iron plates", "steel", "rivets"	Norway	Chr. Thaulor & Sön
	"2 sets wagon wheels", "ropeway pulleys", "saw blades", "3 oil cans", "files", "carborundum"	Norway	C. S. Christensen
	"ore trays", "cotton waste", "nails, asphalt roof paper", wire", "spades"	Norway	E. A. Smith
	"varnish"	Norway	E. D. Mogstad
	"100 m. electric wire and small isolators"	Norway	E. Fjeldseth
	"3 bar. Cement, 2 bar. tagpix", "ochre", "roof tar", "weighing machine", "wire rope"	Norway	Emil Grönning
	"ropeline renewals", "staples", "screws", "6 pump valves", "washers for ropeline tubs","machinery pump - new body"	Norway	G. Hartmann
	"repair of house"	Norway	H. Kunig
	"painting house"	Norway	Ingebrigt Hansen
	"6 files", "50 flanges"	Norway	Gustaf Aspelin
	"repairs"	Norway	Jacob Digre

(*continued*)

Year	Technology	Country of origin	Supplier
	"furniture"	Norway	Jacob Matteson
	"honorarium-electric engineer"	Norway	John Andersen
	"2 forks and 2 rakes"	Norway	John Rotan
	"double sleigh"	Norway	John Unsgaard
	"chains", "varnish, ochre, rope", "store"	Norway	John Öien
	"Guldal Mine - 2nd instalment"	Norway	John Östeng
	"repair of water, piping"	Norway	K. Lund
	"24 dry batteries"	Norway	Möller and Grönning
	"mining level - repair"	Norway	M. Svensen
	"repair of house"	Norway	N. Christensen
	"belting"	Norway	Norske Remfabrik
	"surveyor"	Norway	O. E. Aalen
	"store and printing and stationery", "store, household goods", "store, stationery, furniture", "store, kitchen"	Norway	Ola Hage
	"inspection of electric plant (government)"	Norway	Olaf Sandvold
	"tarpaulins"	Norway	Oure & Stene
	"cutting and crushing machine"	Norway	S. H. Lundh H.
	unknown	Norway	Sigurd Stave
	"oil cleaner and filter"	?	Storm Martens
	"grates (ovn)", "ovn door"	Norway	Trondhjems Værksteds Udsalg
	"machinery-spur wheel for pumps"	Norway	V. Lowener
	"200 isolators", "100 m. vulcanised wire"	Norway	Winger Elekt. Værksted
	"spur wheels"	Norway	Örens Mek. Værksted
	"vacuum oil: 2 bar, dynamo oil"	?	?

(*continued*)

Year	Technology	Country of origin	Supplier
1914	"construction", "ropeways"	Norway	A. E. Gildseth
	"tools, packing, files", "3 pulleys"	Norway	A. E. Olsen
	"writing paper"	Norway	A. Holbæk Eriksen & Co
	"planned and grooved boards"	Norway	A. J. Nilsson
	"copying paper"	Norway	A. H. Eriksen & Co
	"government inspection of electric plant"	Norway	A. Sandahl
	"soldering lamp"	Norway	Albert E. Olsen
	"fuse"	Norway	Andersen & Ødegaard
	"pulley and tools" "brass netting", "rack saus", "springs", "India rubber tubing", "block pulleys"	Norway	A/S Hynnes Maskinforretning
	"trugs and krafsere"	Norway	A/S Arthur Motzfeldt
	"electric materials", "glow lamp and shades"	Germany (Norwegian subsidiary)	A/S Siemens-Schuckert
	"belting"	Norway	A/S Den Norske Remfabrik
	"electric material", "motor repair"	Switzerland (Norwegian subsidiary)	Brown Boveri
	"krafsere"	Norway	Chr. Mynthe
	"glass", "screws", "washer", "iron", "nails", "rail spikes"	Norway	Chr. Thaulor & Sön
	"bands for typewriter"	Norway	Chr. Johnsen
	"pulleys"	Norway	C. S. Christensen
	"belting"	Norway	Den Norske Remfabrik
	"asphalt paper", "pickaxes", "shafts" "paint", "sheet iron", "rail spikes", "spades", "nails", "iron sheets"	Norway	Emil Grönning
	"1 stolling"	Norway	E. Kroken
	"(construction) machinery"	Norway	Edward Kvam
	"fuse & nails"	Norway	E. A. Smith
	"telephone"	Norway	Elektrisk Bureau
	"2 plumbobs"	Norway	Gundersen & Løken

(*continued*)

Year	Technology	Country of origin	Supplier
	"pump valves", "dices"	Norway	G. Hartmann
	"ropeway erection"	Norway	Gustav Thorkildsen
	"chemical manuals"	Norway	Gunnar Birkeland
	"erecting ropeway"	Norway	Ing. Thorkildsen
	"building over ropeway-dressing plant", "construction-tubing over surface", "ropeway", "cutting"	Norway	Jens J. Grönli
	"repair of saw blades"	Norway	Jacob Digre
	"tubing"	Norway	K. Lund
	"erection of oil regulator"	Norway	Kværner Brug
	"electric lamps"	Norway	Möller & Grönning
	"2 nipper and 1 tap"	Norway	M. Engrelius & Sön
	"repair of wagons"	Norway	M. Hjelde
	"paint", "screws", "white lead and chalk", "emery paper"	Norway	M. Engzelius & Sön
	"2 stools and basket"	Norway	Norsk Husflids Venner
	"stove pipes"	Norway	O. A. Moxness
	"manilla rope"	Norway	P. O. Jensen
	"photograph material"	Norway	S. Engan
	"wire netting", "rope hooks"	Norway	Sigurd Stave
	"horseshoes"	Norway	Ragnvald Lillevold
	"spur wheels and ovens" "repair (unknown)" "1 pulley and swings"	Norway	Trondhjems Mek. Værksted
	"2 form"	Norway	Trondhjems Landsfængsel Utsalg
	"sundry goods"	Norway	Ragnvold Lillevold
	"office goods"	Norway	Waldemar Janssen
	"2 spur wheels"	Norway	Ørens Værksted
	"china ink", "drawing paper", "pens", "paper plan", "copying paper", "6 books of 500 pages-timebooks"	Norway	Waldemar Janssen
	"fishplates", "iron", "bolts"	?	?

(*continued*)

Year	Technology	Country of origin	Supplier
1919	"screws", "oil cans", "sundries"	Norway	A. E. Olsen
	"stationery"	Norway	A. Holbak Eriksen
	"board"	Norway	A/S Emil Grönning
	"boring tool", "sundry tools"	Norway	A/S E. A. Smith
	"pump-valves"	Norway	A/S G. Hartmann
	"gum-sales"	Norway	Andersen & Enger
	"one air-receiver for plant"	Norway	A/S Meraker Gruber
	"axle", "electric material", "erector (ropeway)"	Switzerland (Norwegian subsidiary)	A/S Norsk Elektrisk & Brown Boveri
	"belts"	Norway	A/S Viking Rem & Pakn. Fabr
	"stationery"	Norway	A/S Waldemar Janssen
	"sundries"	Norway	Aalens Øvdre Forbruksforening
	"forward machine from England in February"	England	Bachke & C.
	"stationery"	Norway	Bruns boghandel
	"repair typewriter"	Norway	Ch. Johnsen
	"carbide"	Norway	Ch. Engzelius & Sön
	"window-glass", "iron", "one iron-girder"	Norway	Chr. Thaulow & Sön
	"carbide", "sundry painting goods"	Norway	Emil Grönning
	"lac"	Norway	E. D. Mogstad
	"electric fittings", "stationery", "calculating tables", "rope", "leather"	Norway	E. Fjelseth
	"roofing"	Norway	E. Grönning
	"tools"	Norway	E. Olsen
	"tools"	Norway	Eidet Handelssamlag

(*continued*)

Year	Technology	Country of origin	Supplier
	"board & sundries"	Norway	Engzelius & Sön
	"books for office"	Norway	F. Bruns Bokhandel
	"foundations, dam. wall"	Norway	Fjeldsjöen Dam. RO. Aune
	"sorting hammers"	Norway	Fr. Melbye & Co
	"sundries"	Norway	Grønning Emil
	"electrical inspection"	Norway	Government
	"repair of unknown"	Norway	Hamar Jernstøberi & Mek. Verksted
	"custom ovens", "transmission tubes", "transmission washing machinery", "cores", "transmission one case bolts"	Norway	H. F. Bachke
	"assays"	Norway	Hornhauer
	"sundry shop machinery"	Norway	Hviles el. Verksted A/S
	"dictionary", "letter files", "drawing material"	Norway	J. Brun
	"contract reconstruction mine"	Norway	J. Gaare
	"curtains for new office"	Norway	J. Matheson
	"office effects"	Norway	Janssen, Waldemar A/S
	"buildings"	Norway	Johan Gaare
	"sundries	Norway	Jongzelius & Søn
	"sundries"	Norway	Killingdal prov. Handel
	"nails"	Norway	Kvam & Gisvold
	"repair of typewriter"	Norway	Lorentz Syvertsen
	"screws", "sundries", "nails", "rope", "sundries"	Norway	M. Engzelius & Sön
	"sundry goods"	Norway	M. Gallus
	"electric materials", "electric battery", "electric sundries"	Norway	Møller & Grønning
	"electrical material"	Norway	Nicolay Beck
	"nitroglycerine: explosives dynamite"	Norway	Nitroglycerincompagniet
	"caps"	Norway	Nordenfs. Sprængstof A/S

(*continued*)

Year	Technology	Country of origin	Supplier
	"assays"	Norway	Norges Tekniske Høiskole
	"diesel motor contract-pump"	Norway	Norsk HB
	"repair at Storvoldsen"	Norway	Norges statsbaner
	"belting"	Norway	Norske Remfabrik
	"sundries"	Norway	Olsen Alb
	"surveying mine"	Norway	Ole A. Aalen
	"wheels and axles"	Norway	Pay & Brick
	"sundry-building goods"	Norway	P. B. Paulsen & Co
	"sundry goods"	Norway	Ragnvald Lillevold
	"sundry"	Norway	R. E. Carr's Ex. Kristiania & Trondheim
	"tubing", "tubes"	Norway	Sefsaas & Co.
	"cement", "tubing"	Norway	Trondhjems Cementsrøberi
	"washers for grizzlies", "cooking apparatus"	Norway	Trond. Mek. Verksted
	unknown	Norway	Trondhjems Elektricitetsverk
	"bolting store"	Norway	The Foldal Copper & Sulphur Co. Ltd
	"stationery"	Norway	Wald. Jansen
	"furniture for mine house", "sundry provisions for mine house", "sundry provisions for mine house", "sundry goods"	?	?

[a]The cash books after 1919 also show a large share of Norwegian suppliers, but the large number of notes and comments makes it too difficult to include more years

Sources: Bede Metal Killingdal, Privatarkiv 107, Statsarkivet in Trondheim: 10 kassabok sept. 1895–okt. 1906; 11 kassabok nov. 1906–okt. 1919; 12 kassabok nov. 1919–des. 1919

Index[1]

[1] Note: Page numbers followed by 'n' refer to notes.

K. Ranestad, *Knowledge-Based Growth in Natural Resource Intensive Economies*, Palgrave Studies in Economic History, https://doi.org/10.1007/978-3-319-96412-6

H

I

J

K

L

M